SAFETY AND RISK IN SOCIETY

NATIONAL LEVEE SAFETY AND LOCALLY OPERATED LEVEES

SAFETY AND RISK IN SOCIETY

Additional books in this series can be found on Nova's website under the Series tab.

Additional E-books in this series can be found on Nova's website under the E-book tab.

NATURAL DISASTER RESEARCH, PREDICTION AND MITIGATION

Additional books in this series can be found on Nova's website under the Series tab.

Additional E-books in this series can be found on Nova's website under the E-book tab.

SAFETY AND RISK IN SOCIETY

NATIONAL LEVEE SAFETY AND LOCALLY OPERATED LEVEES

DAVID MANCUSO
AND
VIVIENNE MASTERSON
EDITORS

Nova Science Publishers, Inc.
New York

For permission to use material from this book please contact us:
Telephone 631-231-7269; Fax 631-231-8175
Web Site: http://www.novapublishers.com

Library of Congress Cataloging-in-Publication Data

National levee safety and locally operated levees / editors, David Mancuso and Vivienne Masterson.
p. cm.
Includes bibliographical references and index.
ISBN 978-1-61470-511-6 (hardcover)
1. Levees--United States--Safety measures. I. Mancuso, David. II. Masterson, Vivienne.
TC533.N38 2011
627'.420289--dc23 2011024428

Published by Nova Science Publishers, Inc. † New York

CONTENTS

PREFACE

This book examines national levee safety with a focus on recommendations and strategic plan implementations of the National Committee on Levee Safety. It is a critical juncture in our nations' history with a burgeoning growth of risk to people and infrastructure as a result of more than 100 years of inattention to levee infrastructure combined with an economy and social fabric that are in a particularly vulnerable state. The current levee safety reality for the United States is stark and uncertain in location, performance and condition of levees and a lack of oversight, technical standards, and effective communication of risks. A look to the future offers two distinct possibilities: one where we continue the status quo and await the certainty of more catastrophes or one where we take reasonable actions and investments in a National Levee Safety Program that turns the tide on risk growth.

Chapter 1- This report contains the recommendations and strategic plan for implementation for a *National Levee Safety Program* from the National Committee on Levee Safety (Committee). The Committee is a diverse group of professionals from federal, state, local/regional governments and the private sector that have worked diligently at representing national interests in levee safety. The report is in response to Title IX, known as the National Levee Safety Act of the Water Resources Development Act of 2007, specifically Section 9003. As a group, the Committee cannot over-emphasize the urgency of these recommendations.

Chapter 2- We are at a critical juncture in our nation's history – risks of loss of life, property damage, and damage to our natural environment behind levees are increasing. The consequences of levee failure could be devastating. The situation is the result of more than 100 years of inattention to, and in some cases neglect of, levee infrastructure, combined with a growing population living in leveed areas and an economy and social fabric that are in a particularly vulnerable state. Due to decades of inattention and underinvestment, there is not a full understanding of where the risk exists and how significant the impacts of failure might be.

Chapter 3- Locally operated levees and the federal programs that assist and accredit them are receiving increasing congressional attention. Congressional authorization of the National Flood Insurance Program (NFIP), managed by the Federal Emergency Management Agency (FEMA), expires on September 30, 2011. The pending reauthorization has increased congressional awareness of the link between the condition of locally operated levees, FEMA's Flood Insurance Rate Maps (FIRMs) and levee accreditation (which determine which NFIP requirements and premiums apply in an area), and programs providing federal disaster assistance for these levees. Congress is considering whether and how to change current programs, federal funding, and the existing division of levee responsibilities. Options are complicated by the desire to promote state, local, and individual decisions and investments that reduce flood risk; concerns about the local costs associated with NFIP purchase and levee accreditation requirements; and consideration of whether to expand federal responsibilities and potential liability. Even though similar issues also exist for some of the federally operated levees, this report focuses on locally operated levees since these dominate the national levee portfolio.

In: National Levee Safety…
Editors: D. Mancuso, V. Masterson
ISBN: 978-1-61470-511-6

Chapter 1

RECOMMENDATIONS FOR A NATIONAL LEVEE SAFETY PROGRAM*

National Committee on Levee Safety

EXECUTIVE SUMMARY

"The Committee worked assiduously from October 2008 to January 2009, evaluating a wide range of technical, policy and regulatory strategies, with a public safety ethic guiding all decisions. We view the report as the beginning—not the final word—in a national dialogue leading to action among a broad range of stakeholders on our shared responsibilities in levee safety and flood risk management. As a group, we cannot over-emphasize the urgency of these recommendations."

Steven L. Stockton, P.E., SES
Chair, National Committee on Levee Safety

This report contains the recommendations and strategic plan for implementation for a *National Levee Safety Program* from the National Committee on Levee Safety (Committee). The Committee is a diverse group of professionals from federal, state, local/regional governments and the private

* This is an edited, reformatted and augmented version of a A Report to Congress from the National Committee on Levee Safety publication, dated January 15, 2009.

sector that have worked diligently at representing national interests in levee safety. The report is in response to Title IX, known as the National Levee Safety Act of the Water Resources Development Act of 2007, specifically Section 9003. As a group, we cannot over-emphasize the urgency of these recommendations.

We are at a critical juncture in our nation's history—a burgeoning growth of risk to people and infrastructure as a result of more than 100 years of inattention to levee infrastructure combined with an economy and social fabric that are in a particularly vulnerable state. The long history of levees in the United States is full of lessons from both successes and failures. The devastating floods of the late 1920s and 1930s brought a long period of unregulated and poorly constructed levees into focus, resulting in the construction of more robust levee systems for and inappropriate association of the 1%-annual-chance (100-year) event as a safety standard. Our relative complacency during the numerous natural events that continued to wreak economic catastrophes in recent decades was shattered in 2005 in New Orleans. It was the catastrophic loss of life associated with Hurricane Katrina that once again refocused the nation and became the catalyst for the National Levee Safety Act and this report.

The current levee safety reality for the United States is stark—uncertainty in location, performance and condition of levees and a lack of oversight, technical standards, and effective communication of risks. A look to the future offers two distinct possibilities: one where we continue the status quo and await the certainty of more catastrophes or one where we take reasonable actions and investments in a *National Levee Safety Program* that turns the tide on risk growth. We strongly recommend the latter.

The Committee's recommendations are prefaced by recognition of a need for a broader national flood risk management approach, the benefits of integrating national dam safety and levee safety programs, and call for leveraging levee safety as a critical first step in a national infrastructure investment. The Committee also recognizes that levee systems commonly share the same space as water conveyance and critical ecosystems and habitats, and that working with these interests is vital in effectively managing flood risks.

The specific recommendations for a *National Levee Safety Program (NLSP)* embrace three main concepts: (1) the need for leadership via a *National Levee Safety Commission (Commission)* that provides for state delegated programs, national technical standards, risk communication, and coordinating environmental and safety concerns; (2)the building of strong levee safety programs in and within all states that in turn provide oversight, regulation, and critical levee safety processes; and (3) a foundation of well-aligned federal agency programs and processes.

The following is a summary of the twenty recommendations:

Comprehensive and Consistent National Leadership

1. Establish a National Levee Safety Commission to provide national leadership and comprehensive and consistent approaches to levee safety including standards, research and development, technical materials and assistance, training, public involvement and education, collaboration on environmental and safety issues, facilitation of the alignment of federal programs and design, delegation and oversight of a delegated program to states.
2. Expand and Maintain the National Levee Database to include a one-time US Army Corps of Engineers (Corps) inventory and inspection of all non-federal levees. Baseline information will be included and maintained in an expanded National Levee Database (NLD) in order that critical safety issues, true costs of good levee stewardship, and the state of individual levees can inform priorities and provide data for needed risk-informed assessments and decision-making.
3. Adopt a Hazard Potential Classification System as a first step in identifying and prioritizing hazard in leveed areas. Due to a lack of data regarding probability of failure, initial classifications should be based solely on consequences in order to assist in setting priorities, criteria, and requirements as the NLSP is being established.
4. Develop and Adopt National Levee Safety Standards that will assist in ensuring that the best engineering practices are available and implemented throughout the nation at all levels of government.
5. Develop Tolerable Risk Guidelines in order to facilitate an understanding of the options to reduce identified risks, how uncertainty affects this understanding, and to better inform levee construction/ enhancement decisions and weigh non-structural alternatives to flood risk management in a risk-informed context.
6. Change "Levee Certification" to "Compliance Determination" to better articulate the intent that "certification" under the National Flood Insurance Program (NFIP) requirements does not constitute a safety

guarantee or warranty. The purpose of this change is to more clearly communicate residual risks of living and working in leveed areas.

7. Subject Levee Certifications (Compliance Determinations) under FEMA's National Flood Insurance Program to Peer Review in order to increase confidence in technical determinations of compliance.
8. Swiftly Address Growing Concerns Regarding Liability for Damages Resulting from Levee Failures through exploration of a range of measures aimed at reducing the potential liability of engineering firms and/or government agencies that perform engineering services for levee systems (e.g. inspections, evaluations, design, construction administration, certification, or flood fighting). Congress should address this liability concern as a first priority in order to help ensure state and local interest in developing levee safety programs, and to prevent much needed levee repairs, rehabilitation and certification from coming to a halt.
9. Develop a Comprehensive National Public Involvement and Education/Awareness Campaign to Communicate Risk and Change Behavior in Leveed Areas as an essential element of levee safety by improving public understanding of the role of levees, associated risks, and individual responsibilities to empower people to make risk-informed choices.
10. Provide Comprehensive Technical Materials and Direct Technical Assistance crucial to the successful implementation of consistent national standards to states, local communities and owner/operators.
11. Develop a National Levee Safety Training Program including a combination of courses, materials, curricula, conferences, and direct assistance resulting in an increase in the level of expertise and knowledge in all aspects of levee safety. This would include the development of curricula and certification requirements for Certified Levee Professional programs.
12. Develop and Implement Measures to More Closely Harmonize Levee Safety Activities with Environmental Protection Requirements to ensure that critical levee operations and maintenance is not delayed and that, where possible without compromising human safety, environmentally-friendly practices and techniques are developed and used.
13. Conduct a Research and Development Program that will continually advance state-of-the-art technologies and practices for levee safety

and conduct critical operations and maintenance activities in as cost-effective and environmentally friendly manner as possible.

Building and Sustaining Levee Safety Programs in All States

14. Design and Delegate Program Responsibilities to States to assist state and local governments in developing effective levee safety programs focused on continual and periodic inspections, emergency evacuation, mitigation, public involvement and risk communication/awareness, etc.
15. Establish a Levee Safety Grant Program to assist states and local communities in developing and maintaing the institutional capacity, necessary expertise, and program framework to quickly initiate and maintain levee safety program activities and requirements (cost shared).
16. Establish the National Levee Rehabilitation, Improvement, and Flood Mitigation Fund to aid in the rehabilitation, improvement or removal of aging or deficient national levee infrastructure. Investment (cost-shared) is recommended to be applied to the combination of activities, both structural and non-structural, that combined, would maximize overall risk reduction and initially be focused in areas with the greatest risk to human safety.

Aligning Existing Federal Programs (Incentives and Disincentives)

17. Explore Potential Incentives and Disincentives for good levee behavior through alignment of existing federal programs.
18. Mandate Purchase of Risk-Based Flood Insurance in Leveed Areas to reduce economic flood damages and increase understanding of communities and individuals that levees do not eliminate risk from flooding.
19. Augment FEMA's Mapping Program to improve risk identification and communication in leveed areas and consolidate critical information about flood risk.
20. Align FEMA's Community Rating System (CRS) to Reward Development of State Levee Safety Programs by providing further

incentives to communities to exceed minimum program requirements and benefit from lower risk-based flood insurance rates to policy holders who live in leveed areas.

The Committee recommends phased strategic implementation as follows:

- *Phase I:* Immediately implement critical Congressional and federal agency actions including legislation establishing a *National Levee Safety Program*, completion of an inventory and initial inspection of all levees, establish a *Coordinating Council on Communications for Levees*, requiring mandatory risk-based flood insurance purchase in leveed areas, and addressing barriers associated with levee liability.
- *Phase II:* A five to seven year period that overlaps Phase I that incentivizes the development of state levee safety programs through the deployment of a *National Levee Safety Code*, training, research and development, technical assistance and materials, start-up grants for states, and funds for rehabilitation and mitigation.
- Phase III: Transition to a steady state future where state and local levee safety activities are sustained through incentives, and encouraged through disincentives such as withholding funds from existing programs. Levee safety decisions will be guided by the completion of Tolerable Risk Guidelines.
- A National Levee Safety Program is a wise investment that moves the country away from a reactive disaster assistance environment to a proactive safety-oriented culture where the general public and governments are informed and able to participate in shared responsibilities of risk management and where levees are reliable. In the post-Katrina environment we have a clear and well-justified call to action. Levee safety deserves a priority focus within national infrastructure needs as levees protect much of the other infrastructure—such as roads, bridges, schools, and water and sewer treatment plants—from frequent flooding. The Committee is encouraged by the question asked by Congress in the Levee Safety Act and the validation provided by the Committee's external review team. We view the report as a beginning, not an end, to addressing the issue of levee safety and eagerly anticipate the continued dialogue and action regarding the recommendations in the report. In the spirit of a good beginning, the Committee will seek additional stakeholder and

agency input through a series of national and regional listening sessions that were beyond the accelerated pace of the report, but are important as one of the next steps in realizing a *National Levee Safety Program*.

Chenango River with water surface elevation near top of project during flood of record. Binghamton, NY. 2006—Courtesy of NYSDEC.

Goals for the National Levee Safety Program Title IX, National Levee Safety Act

1. Ensuring the protection of human life and property by levees through the development of technologically, economically, socially, and environmentally feasible programs and procedures for hazard reduction and mitigation relating to levees.
2. Encouraging use of the best available engineering policies and procedures for levee site investigation, design, construction, operation and maintenance, and emergency preparedness.
3. Encouraging the establishment and implementation of an effective national levee safety program that may be delegated to qualified states for implementation, including identification of incentives and disincentives for state levee safety programs.
4. Ensuring that levees are operated and maintained in accordance with appropriate and protective standards by conducting an inventory and inspection of levees.
5. Developing and supporting public education and awareness projects to increase public acceptance and support of state and national levee safety programs.
6. Building public awareness of the residual risks associated with

living in leveed areas.

7. Developing technical assistance materials for state and national levee safety programs.

8. Developing methods to provide technical assistance relating to levee safety to non-federal entities.
9. Developing technical assistance materials, seminars, and guidelines relating to the physical integrity of levees in the United States.

VISION AND APPROACH

House constructed on top of Black River Levee. Near Pocahontas, AR— Photo by Elmo Webb, PE 3/23/08.

Mission Statement

(from Title IX of the Water Resources Development Act of 2007).

"The committee shall develop recommendations for a National Levee Safety Program, including a strategic plan for implementation of the program."

Vision for Levee Safety in the United States

Vision of the *National Levee Safety Program*—"An involved public and reliable levee systems working as part of an integrated approach to protect people and property from floods."

New Orleans 17th St Canal. New Orleans, LA.

Focus of this Report and its Relationship with the Broader Issue of Flood Risk Management

In developing a strategic plan and recommendations for a *National Levee Safety Program (NLSP)*, the Committee focused on those foundational elements defined in the Levee Safety Act, that support the vision statement, while the broader issues of flood risk management were distinguished from those issues specific to Levee Safety. Main areas of focus were:

- Employing sound technical practices in levee design, construction, operation, inspection, assessment, security, and maintenance
- Ensuring effective public education and awareness of risks involving levees.
- Establishing and maintaining competent levee safety programs and procedures that emphasize the protection of human life

- Implementing feasible governance solutions and incentives that encourage and sustain effective levee safety programs at all levels of government, including basic hazard reduction and mitigation measures related to levees.

Pumping water out of subdivision after levee breech repaired. Pocahontas, AR— Photo by Elmo Webb, PE 3/23/08.

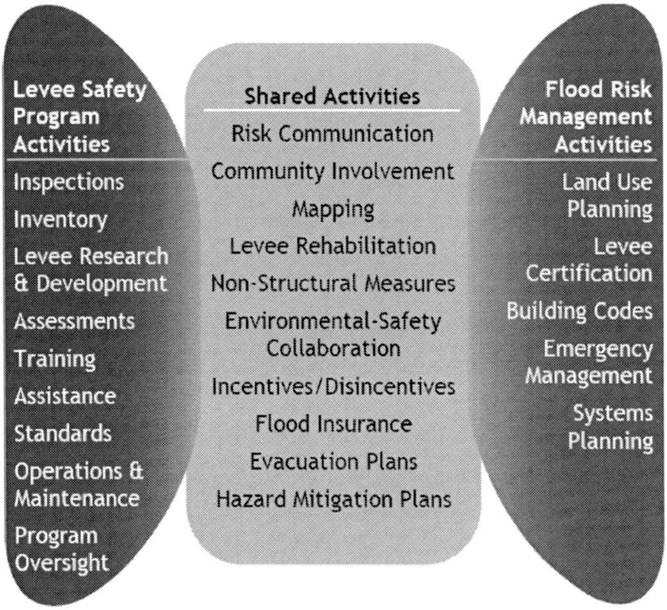

Figure 1. Intersection of Levee Safety And Flood Risk Management Activities with the NCLS Report on a National Levee Safety Program.

In order to achieve our stated purposes, the above four aspects of levee safety were the Committee's primary focus. The Committee explored other goals and connectivity with related flood risk management elements such as insurance, floodplain management, and evacuation, and included recommendations on these issues where they were considered directly related to the scope set out in the Levee Safety Act. Other flood risk management elements, such as land use development and building codes, were less directly related to levee safety and thus deemed outside of the scope of this report. We have endeavored to create a set of recommendations that, as a package, will not only result in a meaningful, comprehensive levee safety program, but place levees in their appropriate place in an overall flood risk management context. After all, in some cases, the safest levee is no levee at all.

Background, Context, and Urgency

The Evolution of Levee Policy in the United States

A Long History

The history of levees in the United States predates even colonization by Europeans. Early Native Americans constructed raised earthen structures along the Ohio and Mississippi Rivers as safe havens from flooding. During the intervening hundreds of years, techniques became more sophisticated, but the general policy of elevating above the flood was still considered effective, if not often employed. From the early days of the country until the 1930s, levee construction around the United States was both sporadic and unsophisticated, and without the benefit of engineering or science practices. Crudely constructed embankments were used to channelize rivers to permit upstream mining (California), protect agriculture and developed areas from riverine flooding (nationwide), transport water for irrigation (West), and provide inland protection along large natural lakes (Florida). These "levees," as we now call them, were prone to breaching from internal defects and overtopping, were essentially unregulated and unmanaged, and often lacked good operation and maintenance practices.

An Early Renaissance Period

The devastation and significant loss of life caused by the great floods on the Mississippi and Ohio Rivers during the late 1920s and 1930s spurred a

Congressional response, ultimately resulting in the Flood Control Acts of 1928 and 1936. These Acts established federal interests in the design and construction of flood control structures such as levees and dams that were to be executed by the Corps at full federal expense. What followed this landmark legislation was the design and construction of thousands of miles of robust levee systems, many providing protection from the "Standard Project Flood"— the largest reasonable flood that could be expected in the basin. Although these levees do not have a level of flood frequency assigned to them, many provided protection from unusual to extreme flooding in the range of 0.2%-annual chance (500-year flood) to 0.1%-annual chance (1,000-year flood). This trend in robust levee construction continued for almost four decades until new national policies began unintentionally encouraging the construction of less protective levee systems.

1928 Flood Control Act

- Established Federal Interest in Flood Control Structures
- Authorized Flood Control Projects on Mississippi River Drainage Basin and Sacramento River
- Other Flood Control Acts and projects to follow

Figure 2. 1928 Flood Contol Act.

Unintended Consequences

In 1968 Congress enacted the National Flood Insurance Program (NFIP).

One of the primary purposes of the NFIP was to address the inability of the public to secure privately backed insurance for economic losses from flooding. Administered by the Department of Homeland Security, the NFIP designated the 1%-annual-chance event (100-year flood) as a special flood hazard area in which those holding federally related mortgages would be required to purchase flood insurance. Never intended to be a safety standard, the 1%-annual-chance event soon became a target design level for many communities as it allowed unrestricted development to continue and provided relief from mandatory flood insurance purchase for homeowners behind levees accredited to meet the 1%-annual-chance event within a relatively economical initial construction cost.

Meanwhile, an interesting parallel was occurring in regards to dams in the United States resulting in a National Dam Safety Program. The destruction and, more significantly, the loss of life as a result of the catastrophic failures of Teton Dam (Idaho, 1976) and Kelly Barnes Dam (Georgia, 1977), resulted in legislation and executive orders for a new national policy initiating the development of the National Dam Safety Program and establishment of the National Dam Safety Review Board, administered and led by FEMA. Today, 49 of 50 states have qualified dam safety programs that provide for public safety through review, regulation, and standards for dams. Unfortunately, there was no correlation between dams and the similar potential that existed for levees.

The 1986 Water Resources Development Act provided new requirements for local cost sharing of flood control projects constructed by the Corps. It also required that lands, easements, rights of way and real estate were to be provided by local sponsors along with an agreement for local sponsors to provide for all operations, maintenance, repair, rehabilitation, and replacement of flood control works. These additional financial burdens on local communities made affordability of new levees and repairs of existing levees an emerging issue and began an unintended shift away from watershed development to individual projects. Combined with the growing and unintended desirability of simply meeting the minimum certification requirements, the affordability concerns resulted in many levee systems over the last 30 years being constructed to provide protection to only the 1%-annual-chance event—a de facto, unintentional, and dangerous adoption of an actuarial standard as a safety standard.

Excerpt from “Risk Analysis and Uncertainty in Flood Damage Reduction Studies”(2000) National Research Council

Why the 100-Year Flood?

The concept of the 100-year flood is central to the National Flood Insurance Program and to many of the Corps’s flood damage reduction activities. Hundreds of government officials administer or work within these flood mitigation and damage reduction programs, to which millions of taxpayer dollars have been devoted. Many consultants are employed in mapping the nation’s 100-year floodplains and scores of university professors analyze the hydrological, statistical and public policy implications of the 100-year flood. Given the economic and social importance of these efforts, one would assume that the selection of the 100-year flood as a defining hydrological event is based on sound scientific and statistical foundations.

Gilbert White, professor emeritus of geography at the University of Colorado, is widely recognized as a leader in promoting sound US flood management strategies. In 1993, Professor White provided an oral interview to Martin Reuss, the Corps of Engineers senior historian. In that interview, White’s response to a question about the selection of the 100-year flood sheds some light on the rational for its selection. Given his knowledge of and experience in the US floodplain management, Gilbert White’s account may be among the better explanations for the prominence of the 100-year flood in US floodplain management and policy.

In response to the question “How do you take into account to so-called catastrophic flood— the once in 100-years flood?”, White stated:

“There was a very interesting development of the notion that there could be a flood of sufficiently low frequency that no effort should be made to cope with it. The Federal Insurance Administration picked one percent [or] a recurrence interval of a hundred years. And some of us were involved in that because we recognized that they initially had to have some figure to use. The one-percent flood was chosen. I think Jim Goddard and TVA colleagues would be considered parties to the crime. With the lack of any other figure, the concept taken from TVA’s “intermediate regional flood” seemed a moderately reasonable figure. We generally use the term “catastrophic flood” for events of much lesser frequency.

This goes back to my earlier criticism of the FIA and it’s determination to cover the country promptly.

In covering the country promptly they established one criterion—the 100-year flood. I think it would have been much more satisfactory if they had not tried to impose a single criterion but had recognized that there could be different criteria for different situations. This could have been practicable administratively even though a federal administrator would say it's far easier, cleaner, to have a single criterion that blankets the country as a whole.

What's the effect of a having criterion of 100 if in doing so a local community is encouraged to regulate any development up to that line and then to say we don't care what happens above that line? We know that in a community like Rapid City the floods were of a lesser frequency than 100 years, and a community ought to be aware of this possibility.

A simplified national policy tended to discourage communities from looking at the flood problem in a community-wide context, considering the whole range of possible floods that would occur.

So I would say that any community ought to be sensitive to the possibility of there being a 500-year flood, or a 1,000-year flood. It should try to consider what it would do in that circumstance, and wherein it could organize its development so that if and when that great event does occur it will have the minimum kind of dislocation."

Gilbert White referred to several risk-related topics addressed in this report. For example, his comment regarding the value of using different criteria for different situations buttresses the Corps's adoption of risk analysis techniques and the abandonment of the levee freeboard principle. As White pointed out, different geographical areas are subject to different levels of flood risk and uncertainty and thereby require different margins of safety. The committee also agrees with Professor White's comments regarding flood hazard preparedness for floods of *all* magnitudes. This committee recommends that rather than focusing on a single event—the 100-year flood—that the Corps examine the risks of flooding from the full range of possible floods.

Complacency Regarding Levees

Riverine flooding on the Mississippi River (1993) and in California (1986 and 1997) spurred additional federal interest in flooding and the role of levees in flood damage reduction and floodplain management when substantial economic damage resulted. Even so, greater catastrophe was only narrowly avoided as most major levee systems protecting heavily urbanized areas held and there was little loss of life. Similarly, several hurricanes along the Florida peninsula (Andrew in 1992, Opal in 1995, Charley, Ivan, Frances, and Jeanne in 2004, and Dennis and Wilma in 2005) and eastern seaboard (Hugo, 1989)resulted in substantial flooding and economic damage but little loss of life. A number of comprehensive and significant reports followed these events, including the "Sharing the Challenge" (Galloway) Report and the Interagency Levee Policy Committee Report (FEMA). Although these reports had well-justified and comprehensive recommendations regarding levees, at that time there was little appetite for creating a levee safety program on a national scale. To date few of the recommended actions have been implemented.

Part of our complacency is related to a misunderstanding of flood risk by decision makers and the general public. Some believe that a 1%-annual-chance (i.e., 100-year) level of flood protection corresponds to a high level of flood protection, perhaps meaning that a flood would not occur for another 100 years. In actuality, a 100-year level of flood protection means that there is a 26% chance of flooding during the 30- year life of a typical mortgage. As shown in the figure on the previous page, even a 200-year level of flood protection corresponds to a 14% chance of flooding over a 30-year period.

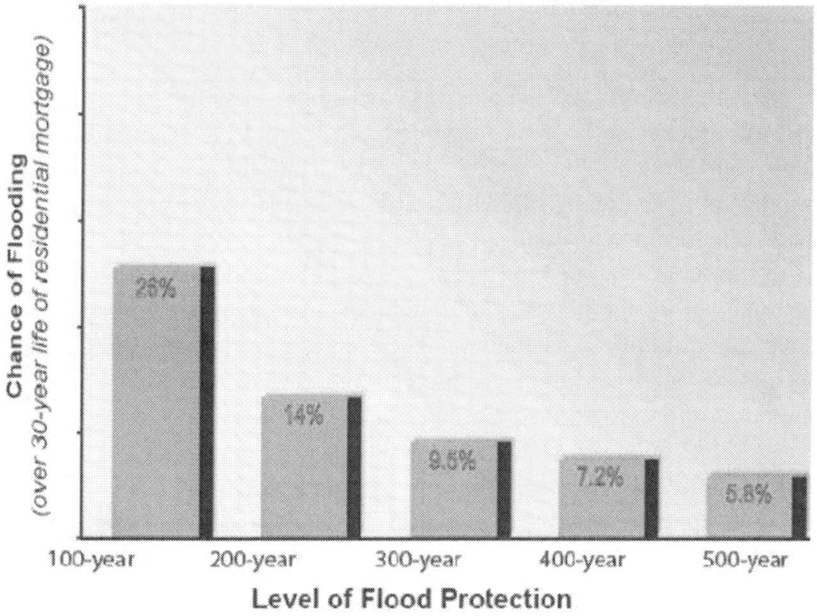

Figure 3. Flood Exposure Behind Levees for Various Levels of Flood Protection (Cumulative over the life of a 30-year mortgage).

Hurricane Katrina. New Orleans, LA. 2005.

These are actually pretty high levels of risk considering that playing one round of Russian Roulette is comparable to a 17% chance of disaster. It is not until we reach a 500-year level of flood protection that the chance of flooding starts getting down to a relatively small chance (i.e., approximately 6% over a 30-year period).

"A flood catastrophe represents a national security issue. Floods especially attack the poor, the disabled and the elderly. They affect our people, our economy, and our environment. How to deal with them has been the subject of many studies over the years and we keep coming back to the same recommendations.

In the future we need to take an approach to flood damage reduction that brings all of the players to the table in a collaborative approach that shares responsibilities and funding. The federal government, acting alone, may not be able to afford new projects but, where it already has been committed to provide protection and where it now provides protection, it has an obligation to provide an appropriate level of protection and to carry out the maintenance necessary to insure system integrity.

Given the tragedies we have seen over the last weeks, the governments and the public must be prepared to take action to 'do it right'—to take recommendations out of the too hard box and move ahead."

Statement of Gerald E. Galloway, PE, PhD

Glenn L. Martin Institute Professor of Engineering University of Maryland, College Park, MD 20742 to the Committee on Transportation and Infrastructure Subcommittee on Water Resources and the Environment US House of Representatives October 27, 2005.

A Wakeup Call

Hurricanes Katrina and Rita (2005) in the Gulf Coast, changed everything. With economic damages estimated to be more than $200 billion dollars and a loss of life of more than 1,800 persons, the role of levees in providing for public safety and flood risk management was again prominently thrust back into the national spotlight. In the midst of an unprecedented federal investment in levee infrastructure and flood insurance in the greater New Orleans area, Congress passed the Water Resources Development Act of 2007—a key element of which was Title IX, also known as the National Levee Safety Act. The Act seeks to develop basic information on federal levees (database, inventory, inspection, and assessments of levees). It also called for this National Committee on Levee Safety. Later in 2008, the flooding and breaching of levees in the Midwest reinforced the sense of urgency. It is the task of this Committee and the purpose of this report to provide recommendations to Congress, including a strategic plan for implementation, for a *National Levee Safety Program*. These tasks require that the current state of levees in the United States—our "Levee Truths"—be fully understood.

Specific Findings

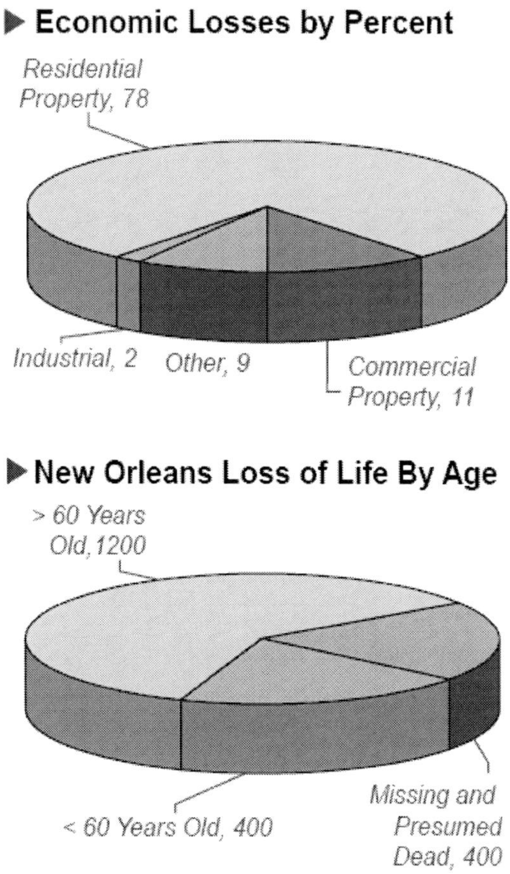

Intergovernmental Performance Evaluation Team Report, 2007.

Figure 4. Consequences of Failure from Hurricane Katrina[1]

1 Since publication of above graph the Louisiana Department of Health and Hospitols placed the final number of confirmed fatalities at 1,810 in all states due to Hurricane Katrina.

The Current State of Levees and Public Safety

An understanding of the current state of levee safety in the United States is necessary if we are to confront the years of neglect and understand the genesis of a new *National Levee Safety Program*.

Levees are now abundant and integral to economic development in many communities in the United States:

- An inventory of the levees under the Corps authorities alone indicates that there are over 2,000 federal levee systems, totaling over 14,000 miles of infrastructure.
- Although the true extent of the national inventory is yet unknown, California has found that the levees designed and constructed by the Corps may represent only 15% of the total levees in the nation—as many as 100,000 miles or more of levees may exist.
- Extrapolating from the federal inventory, it is estimated that tens of millions of people live and work in leveed areas.
- In addition to protecting people and residential property, levees protect much of the civil infrastructure that permits society to function free from frequent flooding, including: roads, railways, bridges, utility systems, water treatment plants, port facilities, critical public service facilities such as fire and police departments and hospitals, sewage treatment plants, refineries and fuel depots, and substantial industry and manufacturing facilities. Levees protect critical infrastructure, facilitating and yielding an economic multiplier effect for communities.
- Although proven beneficial in investment and function, levees have inadvertently increased flood risks in the country by attracting development to the floodplain:
- On average, Corps levee systems currently provide a 6:1 return ratio on flood damages prevented compared to initial costs. Larger, more robust levee systems such as the Mississippi River and Tributaries system provide a 24:1 return ratio on investment. Well-designed, constructed, operated, and maintained levees continue to be economically well-justified federal and non-federal investments.
- Levees can also attract development to the floodplain that would not otherwise be there. The continual growth of population and economic

investment behind levees is now considered the dominant factor in the national flood risk equation (Dr. Pilke, University of Colorado, Wye River Summit, December 2006), outpacing the effects of increased chance of flood occurrence and the degradation of levees. In the 2006 Census, the two fastest growing counties in the United States were St. Bernard Parish and Orleans Parish, both located within the devastated areas of New Orleans, Louisiana.

- The trend for people and communities to locate near rivers and coastal areas is undeniable and will not realistically change in the near future. The link between this natural co-location and the economic welfare of the nation, as witnessed by the effects of Hurricane Katrina on the petrochemical and fishery industries, is obvious. However, as a nation, we have not wisely developed leveed areas in a manner to both realize the benefits of rivers and manage the risks of flooding.

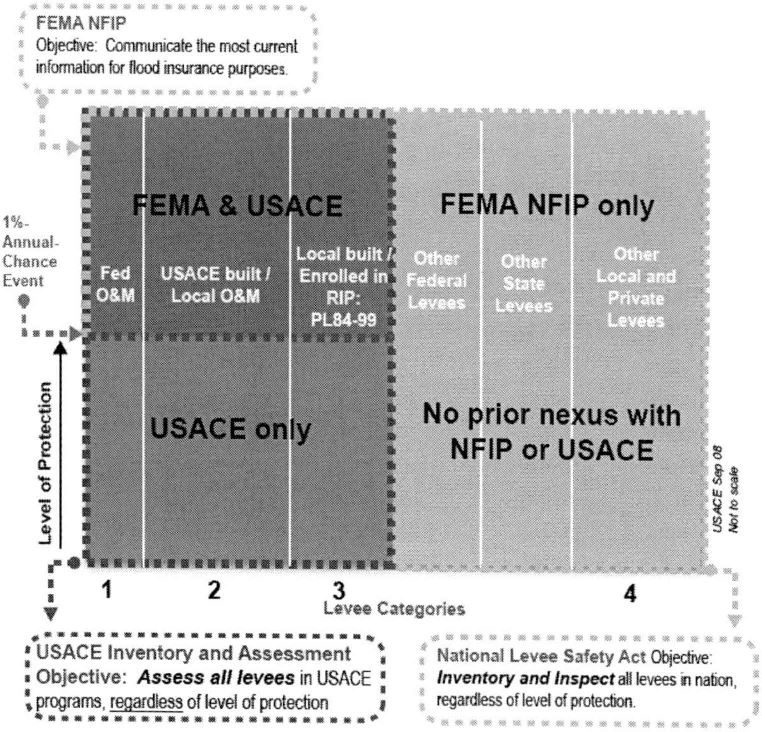

Figure 5. Universe of Levees.

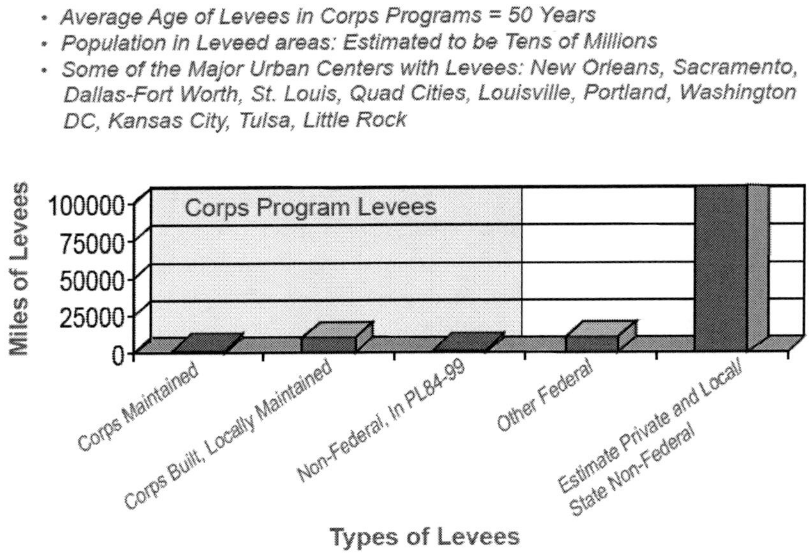

Figure 6. United States Levees at a Glance.

As with all flood control structures, levees only reduce the risk to individuals and structures behind them, they do not eliminate the risk:

- For too long, the partnership of local, state, and federal agencies has allowed the communities in leveed areas to believe that levees—by themselves—make the public safe from flooding. As with virtually any human activity, risks are never eliminated as some residual chance of catastrophe remains and the likelihood of flooding is greater than may be fully appreciated by the public.
- Levees that are poorly designed, constructed, operated or maintained can actually increase risks.
- National discussions have centered on the level of protection offered by levees, and often the risk of living in leveed areas is not articulated. Because of this dialogue on protection, little focus is placed on the measures that the public can take to mitigate their risks.
- Chance and likelihood of flooding remain misunderstood concepts by many. The 1%-annual-chance flood event (i.e., 100-year event) is believed by many to be a highly infrequent event; but in reality, has at least a 26% chance of occurring over the life of a 30-year mortgage for a residence behind a levee. Many Americans located behind 100-year levees do not hesitate to purchase fire insurance for their homes,

but resist the purchase of flood insurance even though the chance of flooding is many times more likely than fire.

The number, location, and condition of all the levees in the United States is currently unknown:

- Knowing the location, condition, owners, operators, and areas protected by levees is fundamental and absolutely necessary to help assure public safety—in fact an inventory of levees is the first step in realizing a national levee safety program. The utility of an accurate inventory also aligns with the concepts of asset management and portfolio management common to good industry practice. Prioritization of activities associated with levees of the highest hazard potential require an accurate inventory of assets.
- By latest count, the approximately 2,000 levee systems just within the Corps program authority account for roughly 14,000 miles of levee infrastructure—this is roughly the same quantity of infrastructure within the entire 84,000+ dams in the National (federal, state, local, private) Inventory of Dams (NID). Therefore, levees by their substantially larger social footprint demand attention exceeding that of dams.
- According to early estimates, non-federal levees may account for an additional 100,000 miles or more of levees nation-wide and other federal agencies like the US Bureau of Reclamation (USBR) are responsible for another 8,000 miles of levee-like structures along canals. Ultimately, levees constitute much more infrastructure that is more integral to communities than do dam infrastructure; but, surprisingly lack the national awareness and safety program focus that benefit dam safety.

Effective flood risk management involves employment of a plethora of strategies, techniques and tools, but in too many instances, levees have been the primary or only tool:

- Evidence suggests that land development controls, building codes, emergency evacuation procedures, flood warning systems, robust levee safety programs, non-structural measures, public education and awareness programs, and flood insurance are all highly effective, but vastly underused tools in flood risk management in the United States.

- Although it is technologically feasible to adequately manage risk through structural means, it is often prohibitively expensive to do so. Consequently, the examples of levees providing high levels of protection—Mississippi River and Tributaries or the Netherlands Coastal Defense—are few.
- The misperception that levees are the single solution to our risk management needs has hindered our ability to achieve a more comprehensive vision of shared flood risk management from being realized and properly embraced by local, regional, and state governments and the individuals that live behind levees.
- Levee systems commonly share the same space as water supply conveyance and critical ecosystems and habitat. As a result, proper management of levee systems must interact and coordinate with these two other important interests. In many cases, this will either place restrictions or create opportunities in maintaining or improving levee systems.
- In general, flood risks cannot be effectively reduced without a significant understanding and employment of non-structural risk reduction techniques.

There is currently no national policy relating to the safety of levees:

- Federal and state agencies have varying policies and criteria concerning many aspects of levee design, construction, operation, and maintenance; but, there are no national policies, standards, or best practices that are comprehensive to the issues of levee safety and that can be adopted broadly by governments at all levels.
- Consequently, the level of protection and robustness of design and construction vary considerably across the country, helping to create a wide-ranging profile of risk exposure, risk understanding, risk levels, and consequently public safety.
- The lack of national standards for critical services in design, construction, and certification of levees. The legal environment—as evidenced in the post-Katrina lawsuits, appears to be making such work increasingly riskier business propositions in comparison to the fees generated. Together, these experiences are effectively reducing the private sector's interest and ability to provide these services.

- Many government officials and the general public have only a limited understanding of levees and the risks associated with them:
- Even competent agencies with large levee inventories such as the Corps or the California Department of Water Resources recognize massive gaps in their knowledge regarding federal levees within their authorities. Such data gaps include subsurface conditions, hydrologic conditions, performance history, design and construction records, inspection data, potential failure modes, modifications, ownership, and the like. Without this information, there is great uncertainty in how reliably the levees will perform in the infrequent and dangerous events during which they are tested. With non-federal levees, anecdotal information suggests that the data gaps are larger and uncertainty is even more critical.
- Uncertainty is a major component of understanding risks—where uncertainty is large, risks are essentially unknown. Without this knowledge, risk awareness is low and risk communication and management is difficult, if not impossible.
- Good decision making relies on quality information. Therefore, major investments in the study and rehabilitation of levees in the United States must be justified by more and better quality information than currently exists.
- Better information on levees will enable more effective public education and awareness of risks. With this information, FEMA's concept of communicating "early, often, and continually" needs to be more thoroughly applied to communicating the risks associated with living in leveed areas.
- Many levees were originally constructed without the benefit of modern engineering techniques and now provide only limited protection to communities:
- The average age of levees within federal levee safety programs is approximately 50 years, and the age of many non-federal levees can be much older—100 years or more. Levee infrastructure has the best practice (engineering codes) physically embedded in them at the time of construction, and in a sense, they become museums of the best practices of the past. In many instances, advancements in the state of the art for engineering and science have been considerable, leaving many levees with features that have serious design, construction, and

operational inadequacies. The costs to repair these levees to the current state of the practice will be enormous.

- Modern engineering practices, such as the use of probabilistic hydrologic modeling, geophysical techniques, potential failure mode analysis, and risk and uncertainty assessments are effective in placing the past practices in context Where these new techniques are applied to older levees, the results clearly indicate that better safety standards and practices are needed.

Levee "Truths"

- Levees are now abundant in many communities in the United States;
- Levees have often inadvertently increased flood risks in the country by attracting development in the floodplain;
- Levees only reduce the risk -they do not eliminate the risk;
- The number and location of all the levees in the United States is currently unknown;
- Levees have too often been the primary tool in flood risk management;
- There is currently no national policy relating to the safety of levees;

- Government officials and the general public often have only a limited understanding of levees and the risks associated with them;
- Many levees were constructed without the benefit of modern engineering and provide only limited protection to communities;
- Many levees originally constructed to protect agricultural fields now protect large urban communities;
- Many urban areas protected by levees, particularly those in deep floodplains, place people who live behind them at an unacceptably high risk. Failure of such levees can result in high loss of life, property damage, and economic losses; and
- The reliability of many levees is commonly not known.

Photo: Chino Canyon Levee. Palm Springs, CA. 2008— Courtesy of Riverside County Flood Control and Water Conservation District.

Palm Canyon, CA— Courtesy of Riverside County Flood Control and Water Conservation District.

Many levees originally constructed to protect agricultural fields now protect large urban communities and the infrastructure they depend on:

- Risk is the product of the chance of the flood event, the likelihood that levees will perform as intended, and the consequences of poor performance. Development in leveed areas—residential, industrial, critical facilities, and civil infrastructure—has resulted in "risk creep"—the steady increase in risk levels over time.
- Federal policies limit the federal investment in levees to the amount that can be economically justified based on existing conditions. Consequently, even levees designed to the full capacity of federal principles and guidelines can soon become inadequate if significant development continues to occur.
- Many levees were planned, designed, and constructed with a specific use and purpose in mind. Other levees lack good engineering practice from inception. In general, protection of higher consequence areas requires more robust engineering standards and levels of protection. Therefore, changes to land development over time and advancements in engineering practices can change levels of public safety needed and required.
- Many urban areas protected by levees, particularly those in deep floodplains, have an unacceptably low level of flood protection and an unacceptably high risk. Failure of such levees can result in high loss of life, property damage, and economic losses.

The reliability of many levees is commonly not known:

- Floods do not respect the political and ownership boundaries by which many levees are managed. Floods exploit system weaknesses across the entire line of protection or system, which may include multiple owners and even infrastructure such as railroad and highway embankments that were not designed for the purpose of flood protection.
- Systems approaches to levee safety demand greater collaboration between levee segment owners and communities.

Safety programs can and should provide improved public safety through the close scrutiny of levee conditions and risks posed, and the communication of those findings to decision makers and affected populations:

- Based on a recent survey of states by the Association of State Dam Safety Officials, only 22 of 50 states had some limited authorities in regulating and overseeing levee safety. None of the states had comprehensive safety programs geared to all of the major components recommended in this report.
- A similar review of federal agencies with responsibilities for levee safety indicates either newly formed programs (US Army Corps of Engineers—2007) or a general lack of rigorous oversight exists (US Bureau of Reclamation, Natural Resource Conservation Service, International Boundary and Water Commission).

Association of State Dam Safety Officials, February 8, 2006

- Does your agency have regulatory authority or responsibility over levees?
  - No: 24 states
  - Yes: 23 states
- If you do not, which agency in your state (if any) does?
  - Most common answer: unknown
  - Misperception that the Corps was responsible
- Describe what types of programs your state has for managing levee safety.
  - Highly varied responses: not regulated to regulated "like dams"
- From your general knowledge, are there levees in your state that cause concern from a safety standpoint?
  - No: 12 states
  - Yes: 25 states
  - Maybe: 10 states

Figure 7. Survey of State Levee Safety Activities.

WHERE WE GO FROM HERE— THE CALL TO ACTION

We can imagine two futures for levee systems and the communities that they help protect: one where we continue the status quo of an uncertain inventory, no national standards, inadequate oversight, lack of risk management, and a degradation of public safety and economic security or; one where we take reasonable actions and make justifiable investments in a *National Levee Safety Program* that help us understand and communicate the risks associated with levees in order that the shared responsibilities of risk reduction activities can be carried out at all levels of government. As a nation, our *Call to Action* is not predicated on *if* the next levee system fails and causes catastrophic damage but *when and where* it fails. The vast numbers of levee systems in the United States combined with their uncertain condition and an increasing flood frequency assure that there will be more such events—it is just a matter of when and where.

Understanding the Future through Risk Concepts

The sense of urgency is most compelling when viewed through the lens of risk:

Components of Risk

Our understanding of future risks associated with levees comes from how the three major components of risk combine: (1) the likelihood of experiencing floods, (2) the likelihood that levee infrastructure and other flood protection measures will perform as intended during these events, and (3) the consequences of poor performance or failure for the protected people, property, and the environment.

Likelihood of Experiencing Floods

Even considering the historical records of the last 100 years or so, engineers and scientists have limited abilities to predict analytically—or accurately extrapolate—the likelihood and flood stage of storms in the future. What we can do with confidence is to show that continued development in the floodplain and within watersheds increases runoff and decreases flood carrying capacity of waterways, thus yielding more frequent and higher flood stages. We can also now conclude that effects of climate change are likely to increase the intensity of coastal and riverine storm events, and thus increase the chance of higher flood stages. In general, we can expect more frequent and higher flood stages in the future to increase the overall risk profile behind levees.

The Likelihood that Levee Infrastructure Will Perform Satisfactorily

Another key element of risk with levees is how well the levee will hold back the anticipated higher and more frequent flood stages. In short, many levees were not built with modern engineering and tend to become less reliable with time. Imagine a 1950s vintage automobile, parked in a driveway since it came off the assembly line, with very limited operation (driving, fueling with leaded gas) and maintenance (oil changes, brake pads) during the intervening years, no improvements related to product recalls or advancements in design (anti-lock brakes, air bags, seat belts, safety glass), no consideration for how the driving environment has changed (speed limits, road surfaces, fuel efficiency) and individual components that have undergone the natural processes of degradation and normal wear-and-tear that come with exposure to the environment. This scenario is the reality within which levees exist—structures that, by and large, lack good maintenance, updates, repairs, and advancements with the state of the art, but that must protect communities from flooding on a moment's notice. So, the trend with the levee performance element of risk is toward lower reliability over time, and thus greater risks.

Consequences of Failure

This element of the risk profile is often both the most dynamic and the dominant factor in the escalation of risk for the protected public. Population growth, and the economic development that comes with it, is not only the fastest growing element of risk but the one that generally has had the least attention and management by governments. In cases where levees are certified for NFIP purposes, development is perversely incentivized through reduced or no requirement for flood insurance and by the potential for governments to build their tax base through development that would not otherwise have been acceptable. Similar to the likelihood of floods and the performance of levees, the growth of consequences is increasing risk over time.

Hurricane Katrina. New Orleans,L.A. 2005.

Tolerable Risk Guidelines

The process that puts all of these components of risk in a societal context and in turn enables better decision making is the use of published tolerable risk guidelines. Although not yet common in levee safety, tolerable risk guidelines have advanced safety engineering and public safety in a number of fields including the airline industry, dam safety programs, transportation industry, and the environmental, food service and medical industries.

WHAT IF WE DON'T TAKE THIS OPPORTUNITY TO ACT?

The other view of the future in regard to levee systems and communities is a continuation of the status quo—no national policies or standards, a lack of oversight and understanding, a lack of education and awareness, and escalating flood losses behind levees. If we are to understand our Call to Action, we must try to imagine the ramifications of this future possibility:

- Envision being surprised by a breaching of a levee system in a major urban area in the United States such as Sacramento, California; St. Louis, Missouri; Dallas, Texas; New Orleans, Louisiana; Hartford, Connecticut; Portland, Oregon; Washington, DC; or Kansas City, Kansas and Missouri. What would be the local and regional effects? What would be the national impacts? International? Where would the people go? How many lives would be lost? How many families would be impacted?
- Now envision these same levee systems as part of the larger systems in society—government, business, the environment, and the social fabric of communities. During, and long after these catastrophes, governments at all levels must operate in a crisis and emergency mode forgoing well-made plans in the process. Businesses—commerce, transportation, insurance, banking, manufacturing, energy—all feel the ripple effect and begin an absorption and redistribution of costs. Environmental effects of contaminated flood waters, destroyed habitat, and second and third order effects of recovery operations increase the stress on already taxed natural systems. And the epicenter of impact—the communities and individuals themselves—struggle to reshape, rebuild, and envision a future for individuals and families at

> just the time when long-term futures are least well-defined and have been most altered. In flooded areas, home values plummet, the single greatest source of personal wealth. One need look no further than the greater New Orleans today to see our future clearly and starkly.

The national response to this all-too-real future will be "Not again! .... How can we be in this position again?" We have the social justification to keep from repeating such disasters—public safety—a key shared responsibility of individuals and all levels of government. We have the economic justifications in terms of flood damages prevented, healthy, striving communities, and the economic benefits/multipliers that come with fixing problems. We have the direction from our national government, and we have the support of our international allies that have already crossed this bridge in developing national safety programs.

Building a different, better future and preventing additional catastrophe and loss is our call to action. A *National Levee Safety Program* is not only a much better offer to the public than the status quo; it is what is expected of us. We must interrupt our patterns of high risk behavior, because it is not only good for "our neighbors" to engage, it is in our self interests to engage. Yes, flood risks are just one source of risks that we as a nation must grapple with; however, it is one for which there is a compelling case for action.

What We Can Do to Secure a Better Future

To have a meaningful chance of slowing and even reducing the levels of risk for communities behind levees it will take a concerted effort to manage all three aspects of the risk equation: likelihood of flooding, levee performance, and consequences of failure. The nation is experiencing a level of flood risk that was not arrived at overnight, but accumulated via a number of practices over the last 100 years or more. It is unreasonable to believe that we can successfully address the causes of our risks in simply a few years—it will take generations of changed behavior and substantial investment. A *National Levee Safety Program* is the first and best step in starting to secure a better future.

A *National Levee Safety Program* begins to address all three elements of risk associated with levees. A comprehensive program of national standards, improved communication, and periodic and continuing safety processes such as an inventory, inspections, and assessments, address the basic data needed to understand and communicate risks. Once this basic information begins to take form, the national program can leverage it to address and prioritize risk reduction activities across all levels of government:

- *Immediate and Short-Term Measures:* consistent interim standards for levee design and construction; more rigorous oversight and review of levee infrastructure by government at all levels; increased public awareness and engagement; evacuation plans; risk-based flood insurance; basic risk mitigation measures in leveed areas; and better understanding and decisions in floodplain development. Results from immediate inventory and inspection activities would inform short-term assessments and rehabilitation of national priority levee systems. States need to assume responsibility for nonfederal levees within their jurisdictions.
- *Long-Term Structural Measures:* a national plan for major rehabilitation, repair, improvement, and/or decommissioning of deficient levee systems.
- *Long-Term Non-Structural Measures:* a national plan for how floodplains are managed that properly balances the desire to place communities near water with the need to better manage flood risks and public safety.
- *Comprehensive, Systems-Based Approaches:* new analytical and decision-making tools that utilize risk-informed applications to evaluate structural and nonstructural measures in concert across entire basins.

Statistics from economic stimulus initiatives indicate that for every $1 billion in infrastructure investment, we create over 47,000 jobs in the economy. So, identifying and fixing the problems in our levee systems not only is a good return on initial investment but creates a multiplier effect in the overall economy.

The American Society of Civil Engineers (ASCE Infrastructure Scorecard) has estimated that the costs to address our nation's failing infrastructure is over $1.6 trillion and increasing. With recent collapses and failures, infrastructure has a national spotlight. Levees are not only part of this infrastructure but form a critical role as flood protection for other infrastructure including roads, railways, bridges, industries, utilities, and water/sewer treatment plants. For this reason, levees and levee safety programs must be an integral element and priority within the larger infrastructure actions.

Recommendations to Congress

The flood risks that this Nation faces are many and varied. During the past twenty years, the recommendation has been made in a number of nationally-commissioned and peer-reviewed reports for a national strategy to address flood risk management.

Closure Installation. Binghamton, NY— Courtesy of NYSDEC.

Even prior to Hurricane Katrina, consistency and collaboration among FEMA and the Corps on flood damage reduction, mitigation, and mapping programs were identified as critical components of a federal flood risk management strategy. Although that effort continues, the loss of life and property due to floods continues to rise and significant deficiencies remain for local and state flood risk management efforts.

While improving levee safety will enhance public safety, the effort will be most effective if it is conducted within the context of a broader national flood risk management program. Levee safety efforts will benefit from a national policy for flood risk management that recognizes the various federal, state, regional, and local responsibilities and functions, provides fiscal support for state and local flood risk management activities, and recognizes state and local governments as the nation's principal flood risk managers.

In presenting this plan, the Committee believes it is important for the reader to understand that while the safety of levees is a significant component of the Nation's approach to flood risk management, it is just that, a component. A *National Levee Safety Program* will be most effective only when coupled with an overall national flood risk management strategy. The Committee recommends that Congress give strong consideration to the development of an overall National Flood Risk Management Strategy, of which the *National Levee Safety Program* would be an integral part.

In addition to the above statement, placing levee safety in an appropriate and useful flood risk management context, the Committee considered the following principles while developing its recommendations:

- Levee safety is a shared responsibility. Responsibilities lie at all levels of government and with persons whose lives and property are located behind levees.
- Our nation's levee problems took generations to build, so it will not be solved overnight. As such, the Committee is recommending a phased approach.
- While levees protect property, infrastructure and economic activity, the Committee has held paramount human health and safety.
- Levees are most effective when managed as physical and political systems, not as individual reaches. We are only as strong as our weakest point.
- Clear attention needs to be brought on issues like: "Who pays?", "Who benefits?", and "Who owns the risk?". If there is an imbalance in these, things will fall apart—the three must be kept in proper tension. Those dealing with land use and those responsible for levee performance must clearly share the risk, the costs and the benefits.
- Commonalities between levee safety and dam safety are many. In order to maximize efficiencies at all levels of government, build upon existing state expertise and provide consistent messages related to multi-hazard risk to the public, all opportunities to integrate the two should be explored; and
- Levees are not only critical public infrastructure, but in many communities protect other critical infrastructure (e.g. roads, bridges, hospitals, wastewater treatment, etc.). Investment in maintaining their reliability should be a national priority.

This recommendations section is organized along the lines of the three major components of what the Committee views as necessary for an effective *National Levee Safety Program*. Under each of these components are specific recommendations:

- *Comprehensive and Consistent, National Leadership*—create a *National Levee Safety Commission* charged with understanding and communicating risks associated with levees, developing national safety standards, facilitating dialogue and research on important levee related topics (e.g. research and development, facilitating dialogue with environmental interests), providing technical materials and assistance to all levels of government, encouraging improved safety measures and programs through grants, and overseeing national and state levee safety program development and implementation activities.
- *Strong Levee Safety Programs in All States*—the cornerstone of an effective *National Levee Safety Program* are effective state programs following a consistent set of national safety standards and mitigation protocols. States are well positioned to provide assistance and oversight to local owner/operators, and coordinate activities in a systems approach among entities within and among states.
- *Alignment of Existing Federal Programs*—in order to ensure that investment in our nation's levees and programs to protect the people who live behind them are effective, all federal programs that impact community and individual behavior in the leveed area should be aligned toward the same goals of risk reduction, developing resilient and reliable levees and protection of human life and property.

Grand Forks, ND.

Before temporary levee protecting subdivision failed. Pocahontas, AR— Photo by Elmo Webb, PE 3/21/08.

Summary of Recommendations for a National Levee Safety Program

The following is an overview of the 20 recommendations described in more detail in this section.

Comprehensive and Consistent National Leadership

1. *Establish a National Levee Safety Commission* to provide national leadership and comprehensive and consistent approaches to levee safety including standards, research and development, technical materials and assistance, training, public involvement and education, facilitation of the alignment of federal programs and design, delegation and oversight of a delegated program to states.
2. *Expand and Maintain the National Levee Database* to include a one-time US Army Corps of Engineers inventory and inspection of all non-federal levees. Baseline information will be included and maintained in an expanded National Levee Database (NLD) in order that critical safety issues, true costs of good levee stewardship, and the

state of individual levees can inform priorities and provide data for needed risk-informed assessments and decision-making.

3. *Adopt a Hazard Potential Classification System* as a first step in identifying and prioritizing hazard in leveed areas. Due to a lack of data regarding probability of failure, initial classifications should be based solely on consequences in order to assist in setting priorities, criteria, and requirements as the NSLP is being established.
4. *Develop and Adopt National Levee Safety Standards* that will assist in ensuring that the best engineering practices are available and implemented throughout the nation at all levels of government.
5. *Develop Tolerable Risk Guidelines* in order to facilitate an understanding of the options to reduce identified risks, how uncertainty affects this understanding, and to better inform levee construction/enhancement decisions and weigh nonstructural alternatives to flood risk management in a risk-informed context.
6. *Change "Levee Certification" to "Compliance Determination"* to better articulate the intent that "certification" under the National Flood Insurance Program (NFIP) requirements does not constitute a safety guarantee or warranty. The purpose of this change is to more clearly communicate residual risks of living and working in leveed areas.
7. *Subject Levee Certifications (Compliance Determinations) under FEMA's National Flood Insurance Program to Peer Review* in order to increase confidence in technical determinations of compliance.
8. *Swiftly Address Growing Concerns Regarding Liability for Damages Resulting from Levee Failures* through exploration of a range of measures aimed at reducing the potential liability of engineering firms and/or government agencies that perform engineering services for levee systems (e.g. inspections, evaluations, design, construction administration, certification, or flood fighting). Congress should address this liability concern as a first priority in order to help ensure state and local interest in developing levee safety programs, and to prevent much needed levee repairs, rehabilitation and certification from coming to a halt.
9. *Develop a Comprehensive National Public Involvement and Education/Awareness Campaign to Communicate Risk and Change Behavior in Leveed Areas* as an essential element of levee safety by improving public understanding of the role of levees, associated risks, and individual responsibilities to empower people to make risk-informed choices.

10. *Provide Comprehensive Technical Materials and Direct Technical Assistance* crucial to the successful implementation of consistent national standards to states, local communities and owner/operators.
11. *Develop a National Levee Safety Training Program* including a combination of courses, materials, curricula, conferences, and direct assistance resulting in an increase in the level of expertise and knowledge in all aspects of levee safety. This would include the development of curricula and certification requirements for a *Certified Levee Professional* program.
12. *Develop and Implement Measures to More Closely Harmonize Levee Safety Activities with Environmental Protection Requirements* to ensure that critical levee operations and maintenance is not delayed and that, where possible without compromising human safety, environmentally-friendly practices and techniques are developed and used.
13. *Conduct a Research and Development Program* that will continually advance state-of-the-art technologies and practices for levee safety and conduct critical operations and maintenance activities in as cost-effective and environmentally-friendly manner as possible.

Building and Sustaining Levee Safety Programs in All States

14. Design and Delegate Program Responsibilities to States to assist states and local governments develop effective levee safety programs focused on continual and periodic inspections, emergency evacuation, mitigation, public involvement and risk communication/awareness, etc.
15. Establish a Levee Safety Grant Program to assist states and local communities develop and maintain the institutional capacity, necessary expertise, and program framework to quickly initiate and maintain levee safety program activities and requirements.
16. Establish the National Levee Rehabilitation, Improvement, and Flood Mitigation Fund to aid in the rehabilitation, improvement or removal of aging or deficient national levee infrastructure. Investment (cost-shared) is recommended to be applied to the combination of activities, both structural and non-structural, that combined, would maximize overall risk reduction and initially be focused in areas with the greatest risk to human safety.

Aligning Existing Federal Programs (Incentives and Disincentives)

17. Explore Potential Incentives and Disincentives for good levee behavior through alignment of existing federal programs.
18. Mandate Purchase of Risk-Based Flood Insurance in Leveed Areas to reduce financial flood damages and increase understanding of communities and individuals that levees do not eliminate risk from flooding.
19. Augment FEMA's Mapping Program to improve risk identification and communication in leveed areas and consolidate critical information about flood risk.
20. Align FEMA's Community Rating System (CRS) to Reward Development of State Levee Safety Programs by providing further incentives to communities to exceed minimum program requirements and benefit from lower risk-based flood insurance rates to individuals who live in leveed areas.

Comprehensive and Consistent National Leadership for Levee Safety

Currently, responsibility for levee safety is assigned in an often uncoordinated and incomplete manner—distributed across all levels of government (federal, state, regional, local) and housed in different agencies and functions within each level of government. This shared and diffuse responsibility impedes development of comprehensive safety policies and programs, impairs ongoing coordination, and prevents a sustained focus on this issue. Effectively addressing levee safety across the country requires a strong, independent, national program drawing on and integrating the diverse expertise from existing agencies at all levels of government and from the private sector.

Recommendation #1: Establish an independent National Levee Safety Commission (Commission) charged with understanding and communicating risks associated with levees, developing national safety standards, facilitating dialogue and research on important levee related topics (e.g. research and development, facilitating dialogue with environmental interests), and providing technical materials and assistance to all levels of government.

Repaired and replanted levee. Dallas, TX – Courtesy of City of Dallas Flood Control District.

Developing Effective Governance for the National Levee Safety Program

The Committee analyzed at a conceptual level how best to govern the NLSP, first considering the "what" of the NLSP, and second "how" the program elements comprising the "what" could best be led and coordinated. The Committee defined the following guiding principles or characteristics as essential:

- Independence to address levee safety holistically, unconstrained by the momentum and priorities of existing programs, and the ability to make politically challenging and unpopular decisions when necessary.
- Leadership for the significant horizontal integration of effort across federal agencies and alignment of their programs, as well as for the vertical integration to achieve strong and balanced participation at all levels of government and in the private sector.
- Organizational capabilities spanning regulatory policy development, program implementation and oversight, grants management; and significant experience in technical, public communications and environmental areas.
- Identifying the most effective governance model to provide for an effective NLSP is neither simple nor obvious. The governing body of the NLSP should have expertise in several areas such as levee engineering, risk mitigation in leveed areas, and administration of grants and incentives, among others. Considering the guiding principles, essential characteristics

- and desired expertise, the Committee developed a governance model dependent on the establishment of a *National Levee Safety Commission* to lead and coordinate the NLSP. Such a governance model provides the strongest organizational basis for the sustained focus and clear accountability needed for levee safety.

Organizational Structure and Duties of the National Levee Safety Commission

The Commission would consist of appointed Commissioners knowledgeable in the fields of water resources and risk management, representing the diversity of skills needed to successfully lead the NLSP including engineering, public communications, program development and oversight, and environment and public safety collaboration. The majority of Commissioners would be selected from state and local government or the private sector, with two of the Commissioners being federal employees, one each appointed by the head of FEMA and the Corps, respectively.

Federal Agencies with Existing Programs and Expertise

The following federal agencies have been identified as having existing programs and/or expertise that would provide a direct benefit in the development and implementation of the National Levee Safety Program.

- US Army Corps of Engineers (Corps)
- Federal Emergency Management Agency (FEMA)
- US Bureau of Reclamation (USBR)
- US Fish and Wildlife Service (USFWS)
- Department of Homeland Security (DHS)
- US Geological Survey (USGS)
- US Environmental Protection Agency (USEPA)
- National Oceanic and Atmospheric Administration (NOAA)
- Federal Energy Regulatory Commission (FERC)
- Housing and Urban Development (HUD)
- International Boundary and Water Commission (IBWC)
- National Resource Conservation Service (NRCS)

The Commissioners' primary duties and responsibilities could include the following:

- Establish and oversee the NLSP, including the program elements and standing advisory committees;
- Review and approve all key regulatory and programmatic changes to the NLSP once established;
- Review and approve delegation of the NLSP to a qualified state or other entity;
- Provide support for delegated programs in facing and overcoming challenges associated with the NLSP development and implementation;
- Review and approve rescission of a delegated program for non-performance;
- Provide periodic recommendations to the President of the United States on the effectiveness of the NLSP including needed authorities, budgets, and coordination with other federal programs;
- Develop and transmit reports to key oversight bodies;
- Conduct periodic evaluations of the NLSP to ensure effectiveness; and
- Understand and communicate risks associated with levees.

To effectively develop, implement, direct, and oversee the NLSP requires that the Commissioners be full-time employees, expected to serve three-year staggered terms, supported by staff consisting of both full-time professionals and additional staff resources drawn from various federal agencies on a temporary and an as-needed basis. This will ensure that the Commission will have sufficient staff resources and expertise as the program is initially developed and launched, and then administered over time. In addition, the Commission will be supported by four standing Advisory Committees comprised of volunteers from all levels of government and the private sector with specific responsibility to advise the Commission on matters related to the NLSP:

- *Delegated Programs Committee* to advise the Commission concerning development and implementation of delegated levee safety programs to qualified states, sustainment of qualified programs at the state level, revocation of delegated programs, management of incentives (including grant programs) and disincentives for state, local and regional programs.

- *Technical Committee* to advise the Commission on matters related to the management of the *National Levee Database*; development and maintenance of the *National Levee Safety Code*, processes for technical assistance to states and training programs; and research and development associated with levee safety.
- *Public Involvement, Education & Awareness Committee* to advise the Commission in the development and fielding of targeted public outreach programs to gather public input, provide education, raise risk awareness, communicate information on delegated programs and track public understanding and behavior changes.
- *Environment & Safety Committee* to advise the Commission on O&M permitting processes for existing projects, coordination of environmental and safety concerns on removal, rehabilitation and new levee projects, and efforts for environment and safety collaboration in leveed areas.

The Commission would establish the size, membership, and specific charter of each standing Advisory Committee, and, as needed, establish additional ad hoc Advisory Committees to address specific topics. Advisory Committee members are anticipated to be voluntary positions drawn from all sectors of government and the private and nonprofit sectors.

Standing up the National Levee Safety Program

The Committee considered two main concepts for governance of the NLSP:

Concept One: Formation of a National Levee Safety Commission

a. Commission established as a new independent federal agency with functional and operational responsibility, and the NLSP placed therein; *or*
b. b.National Levee Safety Program placed in an existing federal agency and the Commission serving as an advisory body to that agency for NLSP duties.

Concept Two: Distribution of the elements of the National Levee Safety Program among various federal agencies without the benefit of a Commission.

Concept 1a: National Levee Safety Commission established as a new independent federal agency

The recommended governance model, a *National Levee Safety Commission*, is represented by Figure 8. The Committee's judgement is that an independent entity, the *National Levee Safety Commission,* would best ensure a strong voice and participation of all key players and provide the appropriate concentrated focus on levee safety and commitment to sustain a comprehensive and robust levee safety program over time. As an independent agency, the Commission would be free from the constraints of many existing competing programs and would be able to provide the critical role of integrating and coordinating across the federal government while providing the single forum for all levels of government to come together to meet their shared responsibilities. For these reasons, the Committee believes that this is the best option and recommends the establishment of a *National Levee Safety Commission* as a new agency to provide leadership in the further development, implementation, and oversight of the NLSP. As work progresses in developing the NLSP, new information and insights will be gathered through expanded stakeholder input, development of the National Levee Database, and additional assessment of the current and potential capabilities of state levee safety programs.

This information will further refine the size and scale needed for the National Levee Safety Commission.

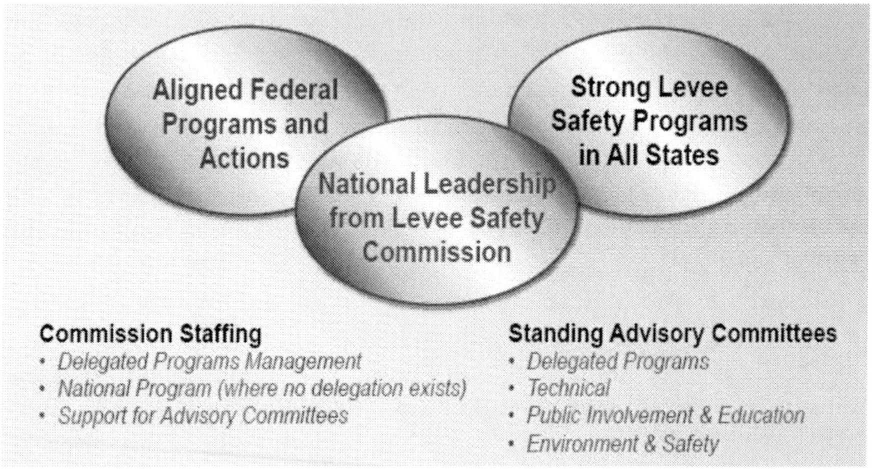

Figure 8. Recommended Governance Structure for National Levee Safety Commission.

Concept 1b: National Levee Safety Program embedded in an existing agency with the Commission as an Advisory Body

The Committee also considered the possibility of embedding the Commission and program in a single existing federal agency, either the Corps or FEMA. While both FEMA and the Corps have strong programmatic involvement with levees and established organizational capabilities and resources, neither is a perfectly ideal home for the program. The governing body of the NLSP should have significant expertise in three important areas: (1) levee engineering, (2) risk mitigation in leveed areas, and (3) administering grants and incentives. While the Corps is expert at the first, FEMA is not, and it would likely take a significant change in culture and possibly organization to develop it there. The Corps and FEMA are both developing expertise in the relatively new field of risk mitigation, but neither has all the expertise needed in this area. FEMA is expert at the third area while the Corps is not, and it would seemingly take a significant institutional change to develop it there. Neither agency has all the expertise needed.

Rather than trying to force such changes and further stretch the resources of these agencies by expanding their already large missions, the Committee believes that it is preferable to utilize the existing expertise from both organizations to support a new, small independent organization that can effectively leverage the resources of both agencies.

In addition, the Committee believes that having the Commission limited to an advisory role within one of these agencies is counter to the realization that levee safety is a shared responsibility across all levels of government needing consensus-based solutions. The Commission, drawing its membership from across all levels of government and having decision-making responsibility on key policy and program activities shared by all affected parties, is critical to the success of the program. The Committee believes that it would be difficult to integrate an independent Commission with such important decision making and oversight authority into the existing operational and management structure of either agency.

Concept 2: National Levee Safety Program responsibilities dispersed among existing agencies without the benefit of a Commission

The Committee also considered whether the various elements of the *National Levee Safety Program* could be effectively distributed among various federal agencies leveraging existing programs and organizations. Such an approach would—if feasible— require the least new resources and potentially accelerate some program elements. The Committee believes that this is not a feasible option for three important reasons: (1) it would not lead to the necessary level of integration and coordination across federal programs; (2) without a Commission, charting and sustaining a long-term program would be difficult; and (3) a critical element to the longterm success of the program, and the primary means for ensuring strong state and local participation in the program is the involvement of state and local representation through the Commission and its standing advisory committees. Additionally, the issues surrounding levees are complex on many levels—addressing technical issues, property rights, liability, and communication of complex concepts of risk to the general public. Further, these issues are largely interdependent. To have an effective levee safety program requires that they be addressed in a singular manner, not through the various lenses of existing agencies where their given authorities and practices differ. Finally, a national levee safety program—with its need for sustained programs over a long term to address the serious risk of relatively rare but catastrophic events—would run the risk of being lost among the numerous other important missions and programs run by these organizations.

Program Responsibilities of the National Levee Safety Commission

The following section includes thirteen additional recommendations describing the major program elements and responsibilities envisioned for the National Levee Safety Commission that will take place at the federal level. The recommendations reflect the Committees' strong belief that a consistent, national voice and approach to levees is needed, but that implementation will only be effective through shared responsibility from all levels of government, citizens who live and work behind levees and the private sector. Program responsibilities include:

- Expand and Maintain the National Levee Database
- Adapt Hazard Potential Classification System and Definitions
- Develop and Adopt National Levee Safety Standards
- Develop Tolerable Risk Guidelines
- Change Term "Levee Certification"to "Compliance Determination"
- Subject Levee Certifications (Compliance Determinations) Under
- FEMA's National Flood Insurance Program to Peer Review.

- Address Growing Concerns Regarding Liability for Damages Resulting from Levee Failures
- Lead Public Involvement and Education/Awareness Campaign to Understand Risk and Change Behavior in Leveed Areas
- Provide Technical Materials, Assistance and Training to States and Communities
- Develop and Implement Measures and Practices to More Closely Harmonize Levee Safety Activities with Environmental Protection Requirements and Principles
- Conduct Research and Development to Support Efficient and More Cost Effective Levee Safety Programs
- Design, Delegate and Oversee Program Responsibilities to States
- Coordinate Federal Agency Activites and Programs

Expand and Maintain the National Levee Database

In order to make good flood risk management investments, we must understand more fully the situation under which we are living—namely the location and condition of our nation's levees. Because watercourses do not respect political boundaries, and levees are best understood in systems, data collection must be conducted in a consistent and comprehensive manner across the nation.

One of the most reliable and inexpensive methods of predicting a levee or levee system performance during a flooding event is to document its past performance. To be meaningful and of greatest use, the NLD must contain all germane information needed to make informed decisions and assessments as to the status and reliability of the Nation's levees and levee systems. Any and all decisions that rely on information contained within the NLD are only as good as the data upon which they are based.

Until we have baseline information, gathered through inspections and post-flood performance data, we will not be able to efficiently or cost-effectively:

- Identify the most critical levee safety issues
- Quantify the nation's risk exposure and true costs of maintaining levees
- Focus priorities for future funding
- Provide data for risk-based assessments

Performance Data That Should Be Collected During and After a Flood Event

- Incidents of seepage and/or boils
- Overtopping
- Stability problems
- Waterside and landside erosion
- Flood-fights
- Breaches
- Partial and near failures
- Evacuations
- Lives lost
- Property damage and estimated costs
- Lawsuits
- Findings regarding any levee incidents
- Weather conditions
- Flood stages
- Flood system operations
- Resources used during flood, including
- flood-fights and evacuations
- National Federal Response

Performance Data that Should Be Collected for Routine O&M

- Burrowing animals
- Excessive vegetation
- Problems with encroachments
- Settlements
- Repairs or modifications
- Piezometric and other data

Recommendation #2: Expand the existing federal National Levee Database (NLD) to include inventory and inspection of federal levees (e.g., federally constructed, non-federally operated and maintained levees) and conduct inventory and inspection of all levees (included in levee definitions) on a periodic cycle, not to exceed 10 years. Data should be incorporated into the NLD.

Specific aspects of this recommendation include:

- Gather levee performance data
- Provision for periodic inventory and inspection updates (initial inventory and inspection should be done by the Corps, but maintained by states on an ongoing basis).

Corps' National Levee Database Upon Which Expansion to Non-Federal Levees Could be Based

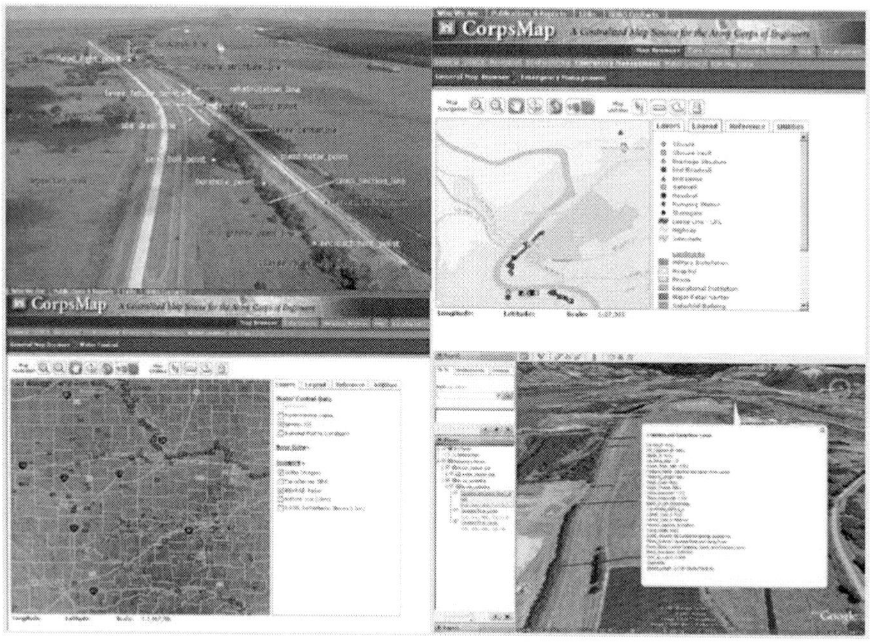

- Development of guidelines related to both the open and limited dissemination of information related to levees.
- Have all state and local governments provide the minimum basic information set out in the National Levee Safety Act.
- Public and private organizations with interest and/or expertise in levee safety should be invited to peer review the NLD and the types of information used in the database.

Hazard Potential Classification	Number of People Potentially Inundated	Number of People Potentially Inundated to Depths ≥ 3 feet	Additional Considerations
High	≥ 10,000	≥ 10,000	Includes areas of consequence where critical life safety infrastructure is at risk (e.g. major hospitals, regional water treatment plants, and major power plants)
Significant	> 1,000	< 10,000	Includes areas of consequence where the number of people potentially inundated is low, but there may be significant potential for large economic impacts or losses
Low	< 1,000	0	-

Figure 9. Hazard Potential Classification.

- Section 9004 of the National Levee Safety Act should be amended to require all state and local agencies to provide data necessary to complete the NLD.

Due to the urgency of this undertaking, Congress should act now to expand current Corps authorities to conduct a one-time inventory and inspection of all the nation's levees (and expand the federal efforts to include performance data). Once the *National Levee Safety Commission* is created, responsibility for maintenance of the NLD and collection of state updates should be conducted by the Commission.

The Corps, in consultation with the Department of Homeland Security's Dam Sector, should establish guidelines to distinguish those portions of the NLD (if any) that, for national security concerns, should not be released to the public.

Develop Hazard Classification System and Definitions

It is expected that both the *National Levee Safety Commission* and delegated programs will need to classify levees by potential hazard, and later by risk, in order to set priorities, criteria, and requirements. The classifications proposed herein, and shown above, are intended for interim use over the next 5 years. During this time, knowledge and lessons learned will be used to develop improved definitions and classifications.

Due to a lack of data at this time regarding probability of failure, definitions and classifications should initially be based solely on consequences of levee failure.

Consequences of levee failure include the following parameters related to the number of people at risk, ability to evacuate (depth of flooding), and property values at risk:

- Population and property at risk within levee flood protection zone
- Depth of flooding—three feet is a common reference where children and the elderly may drown, and evacuation by car or truck is prohibited
- Area and facilities within levee flood protection zone

Height of levee Classifications endeavor, to the extent practicable, to use parameters and definitions consistent with those in use by other agencies (e.g., State of California, FEMA).

- The State of California recently passed flood management
- legislation (Senate Bill 5) and a separate flood bond initiative (Proposition 1E) that define an urban area as having 10,000 people and subject to higher flood protection requirements, and
- also eligible for greater financial assistance from the states.
- FEMA considers shallow flooding in their Special Flood Hazard Areas to be less than three feet.

The proposed three-tier hazard potential classification system shown above is relatively simple, easily understood and quantifiable. It is intentionally set up to parallel the definitions established for the National Dam Safety Program.

Recommendation #3: The Committee recommends that the following levee definitions and preceeding Hazard Potential Classifications be adopted on an interim basis for use with both the national and state levee safety programs. It further recommends that they revised after five years.

Clarifications of Hazard Potential Classification

- Classifications are also intended to include areas of consequence where critical life safety infrastructure is at risk (e.g., major hospitals, regional water treatment plants, and major power plants).
- Also includes areas of consequence where the number of people potentially inundated is low, but there may be significant potential for large economic impacts or losses.
- The area of consequence that establishes the limits for estimating potential hazards should correspond to the elevation of the top of a flood control levee. For canal structures, the area will initially need to be estimated by judgment taking into account the potential volume

that could be discharged by the canal and looking at developed structures within the potential discharge area/drainage.

Levee and Canal Structure Definitions

Levee

A manmade barrier (embankment, floodwall, or structure) along a water course constructed for the primary purpose to provide hurricane, storm, and flood protection relating to seasonal high water, storm surges, precipitation, and other weather events; and that normally is subject to water loading for only a few days or weeks during a year.

Levees may also be embankments, floodwalls, and structures that provide flood protection to lands below sea level and other lowlands and that may be subject to water loading for much, if not all, portions of the year, but that do not constitute barriers across water courses or constrain water along canals.

This levee definition does not apply to shore line protection or river bank protection systems such as revetments, barrier islands, etc.

Levee Feature

A levee feature is a structure that is critical to the functioning of a levee. Examples include embankment sections, floodwall sections, closure structures, pumping stations, interior drainage works, and flood damage reduction channels.

Levee Segment

A levee segment is a discrete portion of a levee system that is owned, operated and maintained by a single entity, or discrete set of entities. A levee segment may have one or more levee features.

Levee System

A levee system comprises one or more levee segments and other features that collectively provide flood damage reduction to a defined area. Failure of one feature within a levee system may constitute failure of the entire system. The levee system is inclusive of all features that are interconnected and necessary to ensure protection of the associated separable floodplain. These levee features may consist of embankment sections, floodwall sections, closure structures, pumping stations, interior drainage works, and flood damage reduction channels. Levee systems include all flood, storm, and hurricane damage reduction systems with any of the major levee features listed above.

Highway and railroad embankments can be considered to be levees only if they are performing as part of a flood control system. While such structures should be considered as part of the levee system, similar to topography, they should be included only to the extent that such structures actually provide some level of flood protection.

Canal Structure

An embankment, wall, or structure along a manmade canal or watercourse that constrains water flows and is subject to frequent water loadings, but that does not constitute a barrier across a watercourse.

Note: Congress included in its direction under Section 9003(2) of the Levee Safety Act that canal structures be considered as levees by this Committee—"[t]he term [levee] includes structures along canals that constrain water flows and are subject to more frequent water loadings..." The Committee strongly agrees they be included for reasons of public safety. Canal structures share with levees many risk and safety characteristics. When many canals were constructed, they were located generally in rural areas, where the major impact of canal failure was the loss of project benefits. With increased urbanization occurring on lands below many canals, significant loss of life and economic damage can now result from failure. To date, many canal operating entities and federal agencies that oversee canals have not independently addressed this problem, and will be important partners in efforts to identify and manage the risk of loss of life and property in canal and levee structure failures. Their inclusion will help assure that national efforts to manage this risk are comprehensive, coordinated and effective.

Unless otherwise stated herin, throughout this report the term "levee" refers to a levee system inclusive of canal structures as defined above.

Unless otherwise stated herin, throughout this report the term "levee" refers to a levee system inclusive of canal structures as defined above.

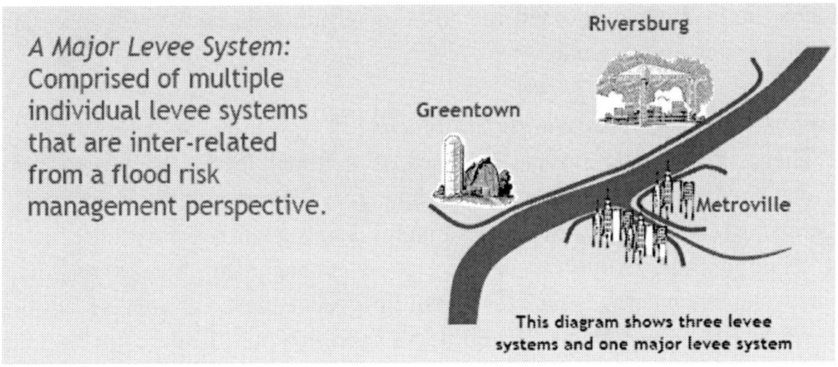

Figure 10. Definition of "System" and "Major System".

National Levee Safety Program Levees

Levees and canal structures should be exempt from regulation under the NLSP if they meet the following conditions:

- A canal constructed completely within natural ground without any manmade structure such as an embankment or retaining wall to retain water and/or where water is retained only by natural ground.
- Highway and railroad embankments that are not functioning as part of a flood control system.
- The levee or canal structure meets *all* of the following criteria:

Not part of a federal flood control project,
and
Not an accredited levee by FEMA,
and
Not greater than 3 feet high,
and
Not protect a population greater than 50 people,
and
Not protect an area greater than 1,000 acres.

Further, in order to avoid duplicative regulations, the Committee considers canals already regulated by the federal government (e.g., power canal regulated by the Federal Energy Regulatory Commission that are subject to dam safety standards) to comply with the NLSP, provided that applied federal safety criteria meet or exceed the to-be-determined interim procedures and National Levee Safety Code.

Develop National Levee Safety Standards

There is currently no uniform set of national levee safety standards. Various agencies use different (or non specific) criteria, making it difficult to understand levee safety across jurisdictions and sometimes creating conflict. For example, the Corps' levee vegetation management memoranda have created major concerns across the nation, especially in California—a conflict that would not have surfaced if well-understood national standards existed and were enforced. Having a uniform set of policies, procedures, standards, and criteria for levee maintenance developed with input from all levels of government, together with input from academia and the private sector, will help establish a common set of expectations across the nation.

Develop Procedures for Three Types of Structures

- Levees that are embankments and floodwalls that have the primary
- purpose to provide hurricane, storm, and flood protection relating to seasonal high water and storm surges, and that normally are subject water loading for only a few days or weeks during a year.
- Embankments and floodwalls that provide flood protection to lands below sea level and other lowlands and that may be subject to water loading for much, if not all, portions of the year, but that do not constitute barriers across water courses, or constrain water along canals.
- Embankments and floodwalls that constrain water along canals, including water supply and power canals.

Engineering Activities Recommended for Inclusion in the Interim Procedures

- Levee Inspections
- Geotechnical explorations
- Site characterizations
- Geotechnical evaluations and analyses
- Hydrologic and hydraulic analyses
- Structural analyses

- Seismic evaluations
- Mechanical/Electrical components
- Levee penetrations (e.g., pipelines)
- Design guidelines and specifications
- Construction administration and
- inspection
- O&M (incl. vegetation management)
- Encroachments
- Security
- Risk analysis
- Levee fragility analysis
- Performance instrumentation
- Residual risk
- Emergency preparedness and response
- Emergency Action Plans
- Flood warning systems
- Flood fighting
- Performance documentation
- Interim risk reduction measures
- Evacuation
- Mapping and risk notification
- Surveys

Further, the development and use of national levee safety standards would provide the private sector with a nationally recognized set of standards that, if applied correctly with appropriate judgment, could help establish a standard of care and probably help reduce the exposure of public agencies and private engineering firms to litigation (see later section for a more in-depth discussion of this critical topic).

Currently, the best documented and available sets of engineering policies, procedures, standards, and criteria related to levees and canal structures are those developed and maintained by the Corps and the US Bureau of Reclamation. Using these as a basis upon which to develop both interim procedures, and eventually the *National Levee Safety Code*, together with the opportunity to update them with input from state, local, academic, and private sector entities, represents the most expedient way to establish well-crafted and accepted policies and procedures for levees and canal structures.

Recommendation #4: Develop and adopt a set of National Levee Safety Standards for common, uniform use by all federal, state and local agencies. The national standards should incorporate engineering policies, procedures, standards, and criteria for a range of levee types, canal structures, and related facilities and features. We recommend that interim products and procedures be adopted by all pertinent federal agencies and used as guidelines by non-federal entities until final standards are developed and adopted by both national and state levee safety programs.

Step One (within 1 year): Develop Interim Guidelines: Under the authority of the NLSP, the Commission should contract with the International Code Council (ICC) to develop *Interim National Levee Engineering Guidelines* (including policies, procedures, standards, and criteria) for levees, canal structures, and related facilities and features using the ICC code development process. This governmental consensus process meets the principles defined in OMB Circular A-119, Federal Participation on the Development and Use of Voluntary Consensus Standards and in Conformity Assessment Activities and Public Law 104-113 National Technology Transfer and Advancement Act of 1995. This expert process is designed specifically to protect the health, safety and welfare of people. It is anticipated that interim guidelines would be based in part on on existing Corps policies, procedures, and criteria for levees and on USBR policies, procedures, and criteria for canal structures as modified through the ICC code development process.

Step Two (within five years): Develop and adopt National Levee Safety Code. The National Levee Safety Commission would again contract with the ICC to take the guidlines developed in step one and further develop them into a National Code.

- The best available practices from other countries should be considered in developing standards, along with lessons learned from using the interim procedures.
- Policies, procedures, standards, and criteria should be linked to *Levee Hazard Potential Classifications* for potential hazard and should incorporate concepts of tolerable risk.
- National procedures, standards, and criteria should be updated every 10 years, or more frequently.

Federal legislation should be passed requiring that all federal agencies and all state levee safety programs adopt the *National Levee Safety Code* once it becomes available. Local flood control agencies participating in either a state levee safety program to adopt the National Levee Safety Code.

Levee damaged due to overtopping. Hurricane Katrina. St. Bernard Parish, LA. August 2005.

Develop Tolerable Risk Guidelines

In order for the nation to better understand the risk associated with living behind a certain levee, more sophisticated approaches are needed. Tolerable risk guidelines are needed to: 1) better enable us to prioritize our public investment at the areas where not only there is a possibility for high consequences, but also where the probability of failure is high; 2) improve citizen and government knowledge and understanding regarding the benefits of mitigation activities; and 3) enhance the public debate regarding the true benefits and costs of flood risk mitigation alternatives.

Because people derive benefits from living in places with high flood risk and demographic trends predict additional influx into the floodplain and coastal areas, we must have tools to help us weigh those risks. We must ask ourselves the following question. *How much protection is reasonable to provide populations against the risk of property damage or personal injury due to flooding?* We can approach this question using a variety of methods:

- Economic calculations on the value of a statistical life saved;
- People's willingness-to-pay to reduce risk;
- State preferences; and
- Risks that people willingly accept.

Tolerable risk methodology can help us better tailor our approaches to investments made and benefits accrued in the levee context. A tolerable risk is one that "society can live with so as to secure certain net benefits." It is a risk that may not be broadly acceptable, and is not necessarily negligible; it is a risk that should be kept under review and reduced if and as possible, but it can be tolerated because of the concomitant benefits. In contrast, intolerable risks are those "so large that nobody should be exposed to [them] and thus risk reduction should be undertaken without regard to cost." *(Reducing Risks, Protecting People: HSE's Decision Making Process (2001), UK Health and Safety Executive, London: HMSO, p. 27)*

Recommendation #5: The National Levee Safety Commission should work with its Standing Technical Committee to develop National Tolerable Risk Guidelines for levees and structures along canals.

Because tolerable risk expertise is so specific, the Commission should:

- Assemble a panel of international renowned experts knowledgeable of tolerable risk concepts with the purpose to develop *National Tolerable Risk Guidelines for Levees and Structures Along Canals.*
- Conduct a peer review of the panel's recommendations by an equally renowned group of experts.
- Enact new federal legislation with requirements for incorporating *National Tolerable Risk Guidelines for Levees and Structures Along Canals.*

Top and Bottom Photos: Levee reconstruction post Hurricane Katrina. St. Bernard Parish, LA. August 2005.

A Primer on Tolerable Risk Guidelines and their Application to our Nation's Levees

What Are Tolerable Risk Guidelines?

Tolerable Risk Guidelines (TRG) are an improved methodology for decision making that enables investors to understand how the infrastructure-related risks for a specific system or portfolio of systems compares to what society and engineering practice deem to be tolerable. The use of TRG not only enables one to put risk in this broader context, but facilitates an understanding of the options to reduce that risk, how uncertainty effects this understanding, and how well justified are the ultimate decisions in order to gain broad stakeholder support. Two common misconceptions about TRG that should be recognized up front:

- TRG do not replace traditional engineering standards, they compliment them by putting considerations such as factors of safety, design approaches, and construction techniques into a consistent context in which to evaluate.
- TRG are not a simple numerical solution, they require the judgment of experienced engineers and scientist to have meaning and support confident, well-justified decisions.
- TRG inform decisions on both structual and non-structural remediation alternatives.

Definition of Tolerable Risk

- Risks society is willing to live with so as to secure certain benefits,
- Risks society does not regard as negligible or something it might ignore,
- Risks that society is confident that are being properly controlled by the owner, and
- Risks the owner keeps under review and reduces still further if and as practicable.

Citation: ANCOLD, 2003.

"As Low as Reasonably Practical"

- The "as-low-as-reasonably-practicable" (ALARP) considerations include a way to address efficiency aspects in both individual and societal tolerable risk guidelines.
- The ALARP consideration states that risks lower than the tolerable risk limit are tolerable only if further risk reduction is impracticable or if the cost is grossly disproportional to the risk reduction. (Adapted from ICOLD)
- Determining that ALARP is satisfied is a matter of judgment.

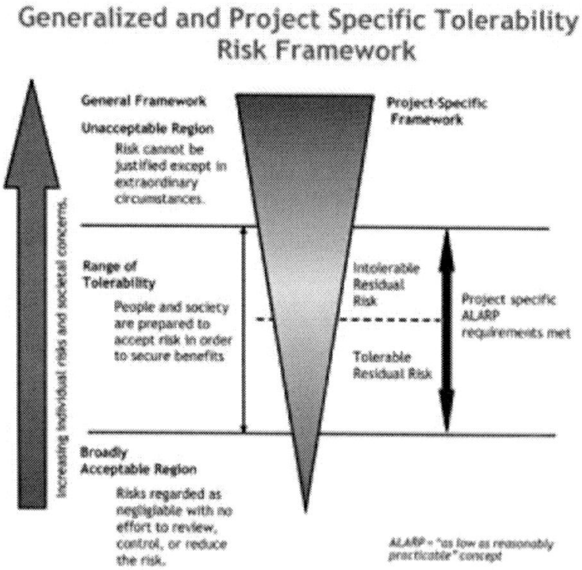

A common International graphical representation of tolerable risk guidelines.

TRG methodology considers how the (1) probability of failure for an element of infrastructure or political system combines with the (2) consequences of failure to create an (3) "annualized consequence risk". Often, the risk is expressed in a loss of life per year metric. All three elements of risk are key metrics that help put the options available to reduce risk into a more logical and organized context.

Some call this process "optioneering"—how engineering options are considered to gain the most cost effective risk reduction. The recognition of the level of knowledge or confidence in the information being evaluated—also known as an uncertainty analysis— is an important aspect of each measure.

A Primer on Tolerable Risk Guidelines and their Application to our Nation's Levees

How Does Tolerable Risk Differ from other Ways of Measuring/Looking at Risk?

By itself, the estimation of risk is significant in determining the priority and relative urgency within a set of conditions and potential actions for remediation, including both structural and non-structural. TRG advances the utility of these risk estimations several significant steps by answering the following questions: (1) what are the limits of tolerability for probability of failure and annualized risk?, (2) how close are the estimated risks to these limits of tolerability? and (3) are there any limitations posed by economic factors or options that further define what is "practicable and achievable" if risks are above a limit of tolerability? For example, it is not just important to know the order (priority) and speed (urgency) at which to take action, it is even more important to know if your suggested actions are understood in a larger context, if they are the best options for reducing risks, if they are well justified, and if they bring conditions to a state of tolerability.

TRG also offer substantially better decision making than traditional standards based decision making as it allows a fair determination of the "worst first" concept, thus facilitating a smart "staged" buy down of risks across a large portfolio.

Central role of TRG in the inter-relationship between risk communication, risk management, and risk assessment.

Why Is Tolerable Risk a Preferred Way of Looking at Levees?

TRG are particularly important when dealing with a massive national portfolio of (on average) 50 year old levees that do not meet most engineering standards. The sheer size and costs of the infrastructure challenges regarding levees will take billions of dollars and decades to realize. Therefore, the order, urgency, method, and justification for rehabilitation action is critical to maintaining credibility and investment support, and for addressing public safety issues in an appropriate manner.

Tolerable Risk: Begin with the End in Mind

- Identify levees that pose
- greatest risk
- To what extent do they need to be modified or risks mitigated? (tolerability)
- Which actions should be taken first? (priority/sequence)
- How do we balance the desire to reduce risk with the availability of resources? (urgency)

Concept of Equity and Efficiency

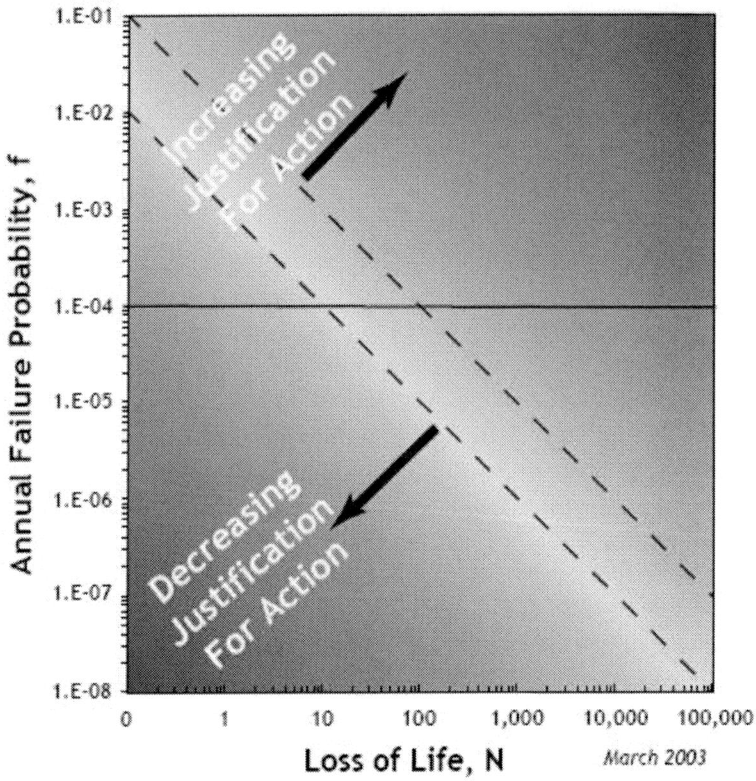

Bureau of Reclamation f-N chart for Displaying Probability of Failure, Life Loss, and Risk Estimates—Portrayal of Risk.

- *Equity*—The principLe, which hoLds that the interests of aLL are to be treated with fairness and that individuaLs and society have the right to be protected (ICOLD);
- *Efficiency*—In reLation to society's use of resources, that principLe, which seeks to gain greatest benefit from the avaiLabLe resources; and
- This Leads to the notion that toLerabLe risk shouLd consider both societal and individual risks as an integraL part of the framework for managing risks.

Removing Barriers Associated with Liability

Under current law, liability could be incurred by state and local government agencies and engineering firms that provide services for levees and other flood control structures and systems. Parties harmed due to levee failures may bring suit against agencies, companies, and individuals involved in levee design, construction and inspection.

The Corps and other federal agencies are afforded immunity from liability of any kind for damage from floods through provisions of the Flood Control Act of 1928. The primary purpose of the immunity provision was to avoid having flood damages added to the very substantial costs of flood control projects that were contemplated. Recently published draft policy states that the Corps will likely no longer certify levees that are not designed, constructed, owned or operated by the Corps. This leaves other government agencies and private engineering firms as the only entities left available to perform this service. These entities are reluctant to provide these services due to a liability potential that, in the case of private engineering firms, far exceeds the fee for services and/ or the entity's financial value. While this issue has been most urgent in the certification realm, some private engineering firms are also no longer willing to provide design and construction services.

Actions should be initiated as soon as possible due to the urgent need for levee engineering services, including certification, across the nation. Many communities and leveed areas have received FEMA notifications that they must recertify their levees within a two year timeframe. In most cases the Corps is not providing this service and have drafted policy that they will not certify non-Corps levees. In reaction to this policy, those seeking certification are looking elsewhere for those services, such as to private engineering firms, and state or local agencies that do not have federal immunity from liability. If this issue is not addressed expediently, it is likely that more private engineering firms and agencies will not offer service where it is most needed.

States, cities, counties, and local districts that begin inspecting levees for which they currently have no responsibility, such as privately owned levees, could be concerned about bringing new liability upon themselves. Inspection of all levees within a state is a key requirement for a state to have a delegated program under the NLSP. To the extent that delegated state programs exceed minimum requirements and take on responsibility for levee permitting, levee construction approval, and operation and maintenance of neglected levees, additional liability concerns may arise to the state and local government entities that undertake these responsibilities. Unless special protections are provided, the liability concerns may be serious enough so as to lead states and local governmental agencies to decline to participate in these actions, or even in the activities necessary to qualify for a delegated levee safety program.

Recommendation #6: Federal agencies should change the term "certification" (such as used in the NFIP) to "compliance determination" to better communicate to policy makers and the public that the determination does not imply a guarantee or warrantee.

Recommendation #7: Levee designs and levee certifications (compliance determinations) for the NFIP should undergo independent peer review.

Recommendation #8: Congress should swiftly address growing concerns regarding liability for damages resulting from levee failures through exploration of a range of measures aimed at reducing the potential liability of engineering firms and/or government agencies that perform engineering services for levee systems (e.g., inspections, evaluations, design, construction administration, certification, or flood fighting). Congress should address this liability concern as a first priority in order to help ensure state and local interest in developing levee safety programs, and to prevent much needed levee repairs, rehabilitation and certification from coming to a halt.

Examples of measures discussed by this Committee include:

a. Limitations on third-party liability for engineering firms providing engineering services for a levee system that might result from a levee failure during a flood event:
   I. Establish that liability following a flood event would only be present if the flood event was equal to or less than the design or rated level of flood protection provided by the levee system;
   II. Establish that the engineering firm would not be liable for decisions (e.g. level of flood protection provided) that are made by other parties (e.g. levee owner or maintaining agencies); and
   III. An engineering firm would be liable only to the extent caused by negligence, recklessness or willful misconduct of the firm.
b. Provisions to limit liability for state and local agencies that sponsor, and then accept, federal flood control projects due to design and construction deficiencies. State and local agencies would benefit from protection against suits alleging damages to persons or property resulting from the construction of the federal flood control facilities.
c. Provisions to limit liability for state and local agencies that, by implementing levee safety programs, provide oversight, funding, or

other levee-related services for non-federal levees unrelated to any provided services.

California Flood Litigation

In the 2003 Paterno decision, the California Court of Appeals found the state liable, by inverse condemnation, for damages incurred by flooded residents as a result of a levee failure along the Yuba River. The *Paterno* decision and others give rise to growing concern of the possible emergence of a strict liability standard being broadly applied in cases of levee failure that result in widespread harm.

In addition, the State of California is now being sued by a railroad for the 2004 levee failure at Jones Tract. The state's role was to provide financial assistance to the local levee district for operation and maintenance and to inspect the resulting work performed by the levee owner, verifying that the funds were spent for their authorized purposes. This experience demonstrates how having any involvement with a levee can create uncertainty about liability.

Lead Public Involvement and Education/Awareness Campaign to Understand Risk and Change Behavior in Leveed Areas

Improving the safety of people who live behind the nation's levees is the top priority of this Committee and should be one of our country's highest priorities. In recent years, thousands of citizens have lost their homes, their livelihood, and in some cases even their lives due to flooding caused by levee failures. Loss of life due to flooding from levee failure can often be attributed to an individual's lack of understanding of the limitations of levee systems and an unrealistic assessment of personal risk. This ultimately results in a failure to take necessary safety measures such as evacuation.

There is an urgent need to raise public awareness of issues related to levees. The public must be educated on the true risks associated with living in leveed areas and how to effectively deal with them. But experience has shown that simply informing individuals rarely affects positive changes in behavior. Success requires both public awareness and public involvement.

Opportunities for public education and public engagement must be provided at all levels of government. Public input is vital to insure that the elements included in a safety program reflect public values. An involved, informed public will be empowered to not only drive their governments to reduce flood risk, but will also take more personal responsibility in buying down that risk. As individuals, they will be better prepared to take risk reduction measures such as purchasing flood insurance, making structural changes to businesses and residences, providing adequate revenue (taxes) for proper levee operations and maintenance and evacuating when required. These measures not only increase public safety and reduce personal loss, but also reduce overall economic loss to the nation thereby lessening a reliance on post-disaster relief.

There are multiple federal state and local agencies (e.g., FEMA, Corps, USBR, local levee owners, etc.) that communicate information about levees and levee safety. Each agency has developed its own message and terminology, resulting in inconsistent and sometimes conflicting messages related to levee safety. This has caused public confusion and frustration. There is no single entity charged with the responsibility of coordinating terminology and message across all the various agencies.

Traditionally, engineers have communicated flooding by using terms such as "100-year level of protection." Such terminology has served to confuse the public and in some cases has led to a false sense of security. Consequences of levee failures are rarely clearly identified. Effective risk communication can only occur when both probability and consequences are included. Numerous governmental and private sector experts have articulated the need to develop a consistent and effective way of communicating flood risk in leveed areas, but to date, no one has developed an effective way of doing so. While levee standards and other technical requirements are most appropriately developed by engineers, a very different set of skills is required to develop effective public education and risk communication programs.

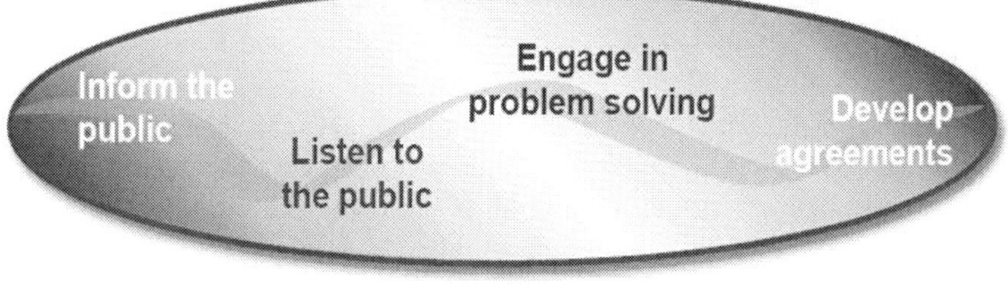

Figure 11. Major Public Involvement Steps.

Each individual living in a leveed area is responsible for mitigating flood risk, particularly when it comes to preserving personal safety and the safety of family members. Levee safety is a shared responsibility and relies on involved, informed, motivated citizens, owner/operators, and governments.

Recommendation #9: Develop a comprehensive national public involvement and awareness/ education program to increase public understanding of the role and limitations of levees, raise awareness of national and state levee safety programs, and effectively communicate risks associated with living in leveed areas.

While the program may be developed at the national level, much of the actual communication will be accomplished at the state and local levels. Public outreach and risk communication activities should be guided by the following general principles:

- Assess the needs and gather input from the public, states, levee owners/operators, local governments and other stakeholders with an interest in public safety in leveed areas. Participation must be actively sought and the program must allow participants to define how they participate. Input must be obtained through realistic and meaningful opportunities. In order to advance shared responsibility, it must be evident to all that contributions from the various groups are being used to influence decisions made by program administrators.
- Ensure consistency of messages across government agencies. A significant benefit of a NLSP is the ability to develop and coordinate consistent terminology and messages across all agencies, enabling the public to better understand levee system-related issues.
- Provide opportunities to educate the public and interested stakeholders on matters pertaining to levee systems and levee safety programs. A national levee safety program is a new concept. The public and interested stakeholders will need to know how the program works, the anticipated benefits of the program, and how they can get involved.
- Ensure that risk communication is clear and consistent. The public involvement and awareness/ education program must emphasize the concept of "risk" and move away from the old terms of "level of protection." The program must include elements to communicate these concepts without technical jargon in a way that people can understand and use to make informed decisions about their lives and property. As conditions in leveed areas change, the level of risk

changes. Therefore, risk information must be updated and communicated on a regular basis.

- Seek to change behavior. Many existing education/awareness efforts only seek to make individuals and governments aware of risk. Merely understanding the risk of living or building behind a levee is not sufficient to protect human life and property. The focus of the NLSP risk communication effort, and the measurement of its success, must be aimed at increasing involvement of individuals, businesses, and governments and persuading them to change their individual and collective behaviors in a manner consistent with increased safety and protection of property.
- Ensure that adequate expertise is available and utilized. We must draw upon the appropriate experts to design, implement and oversee the public involvement and education/awareness program. By involving experts in fields as social marketing, behavioral economics, risk communication, etc., we can better design programs and products to achieve the behavior change we are seeking: an involved public that understands the risks and takes appropriate actions to mitigate them. A high priority element critical for the success of this program is the vocabulary and graphics to describe risk and experts must be engaged to accomplish this. Adequate dedicated funding for this purpose must be provided to ensure success.
- Develop the major components of the public involvement and awareness/education program at the national level for implementation primarily at the state and local levels. Development of the components at the national level will insure a consistent message that can be tailored to meet local needs and serve local audiences. The most effective way to deliver that message is at the local level. The national program should leverage existing best practices in developing its awareness/education program.

Step One (immediately): Lead agencies such as the Corps and FEMA shall establish an ad hoc committee of communication experts from agencies who are currently involved in public education and awareness programs, communicating risks to the public and/or working with the safety of levees. This *Coordinating Council for Communication for Levees* should be housed in FEMA, and work should immediately begin to identify existing programs, link relevant websites, provide public forums to discuss the *National Levee Safety Program* and identify potential Advisory Committee members and experts.

The Council will promote consistency of terminology, messages and approaches across the federal agencies.

Step Two: Establish a *Public Involvement and Education/ Awareness Standing Committee* of the National Levee Safety Commission

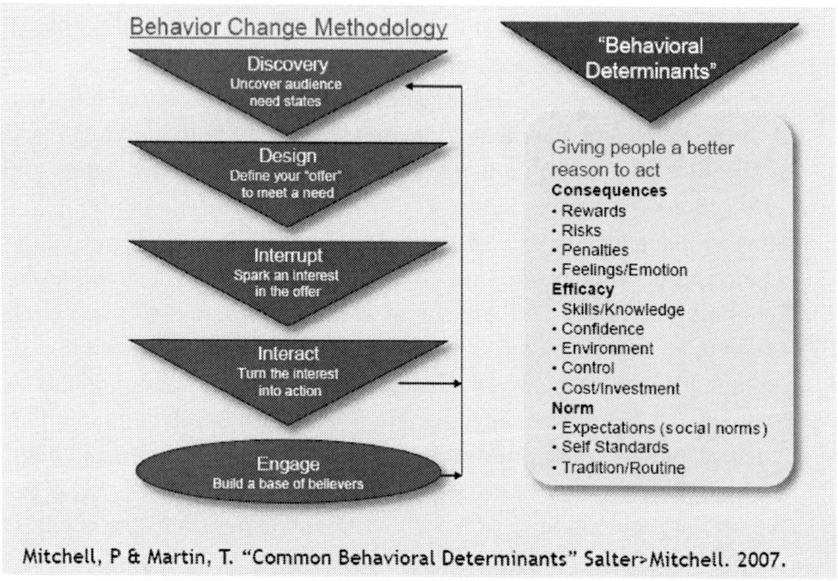

Figure 12. Beyond Risk: you have more than one tool.

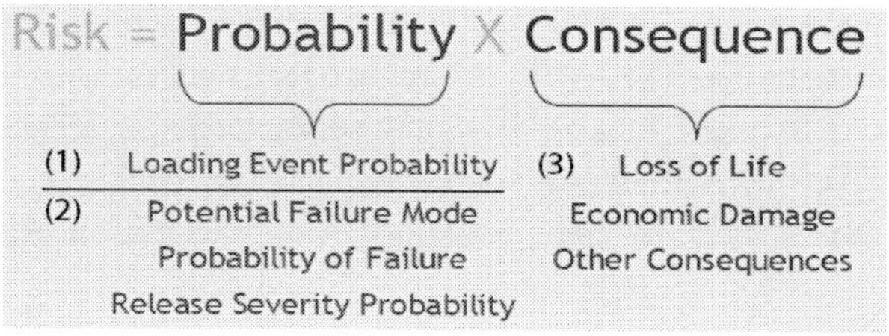

Figure 13. Risk Equation Definitions.

- The standing committees should include federal, state, local and private sector communication experts who will be responsible for the development and implementation of the public involvement and awareness/ education program (9-13 members, ideally). The standing committees should have adequate resources to reach out for specialized expertise as needed for consultation, material generation, peer review, etc.

Example of Regional Risk Maps

These maps are part of a global examination of risk from natural hazards.

- The standing committees should work to ensure better cooperation and consistency between agencies by taking over from the *Coordinating Council for Communication for Levees.*
- The standing committees should establish *national leadership* in all aspects of a comprehensive public involvement and education/ awareness program (e.g. target audiences, messages, tools, materials) as well as develop a rollout/train the trainer implementation. The work of the standing committees will include, but is not limited to, the following elements:

1) An assessment of public understanding and needs that has been developed through professional research and surveys and input from the public. This assessment will tie directly to the goals and measurements established for the program. This element of the program can and should include "listening sessions" across the United States that will increase the profile of the issue of levee safety and get the public interested in the effort.
   The sessions will also provide an excellent database of interested groups and individuals who can later be contacted with additional information.
2) Risk communication vocabulary and components that consistently and clearly explains to the public the risk of living behind levees.

3) Messages, materials and goals aligned with information derived from the assessment and public input, technical recommendations, levee safety policies, and local and state incentives and disincentives.
4) A robust virtual dialogue component including a dynamic, interactive website linked to state and local agencies that can be used for numerous purposes, including continuing the dialogue on levee safety, collaborating, asking questions and getting answers from experts, public discussions, computer simulations, keeping audiences aware of the status of the program in their area, providing communication templates and programs, and housing best communication practices and training tools. This component should also include opportunities for people to interact with the data and to see things in ways that make sense to them such as maps that show inundation levels, videos of homes that have been flooded and other images that will command respect for the damage potential and safety hazard.
5) Materials for use by trainers, government officials, organized by target audience.
6) Training program to teach communication skills and effective use of materials and a program to "train the trainer" to ensure proficiency at the state and local levels.
7) Technical assistance to state and local agencies and private owners.
8) An educational program for school-age children.
9) An annual report to Congress and the public on the state of levee infrastructure, the outcomes of the program that reflect positive changes to our citizens' lives, and the overall efforts and status of the NLSP.
10) Measurement of the effectiveness of public involvement and education and awareness efforts.

Examples of Recommended Materials

- Topical discussion guides (e.g., flood risk management, dam safety, infrastructure)
- Background papers
- *National Levee Safety Program* basics
  - Need for the program
  - Anticipated changes

  - Mechanics/timeline
- Templates
  - How to hold a public workshop,charette, focus group, coffee klatch and advisory group
  - Basics of risk communication — Road signage
  - Developing an evacuation plan
  - How to talk to your community about mitigation
  - Setting up a "Citizen Levee Watch"
- List of potentially interested parties (e.g., civic clubs, COGs, Chambers of Commerce, professional associations)

The following table represents major target audiences, possible sought-after behavior change, and information and tools needed to achieve behavior change. The Committee should consider these, but not be limited by them.

Target Audience	Behavior Change Desired (examples only)	Information and Tools (examples only)
Homeowners	Buy flood insurance elevate/ floodproof home	NFIP information; height of potential flooding; information on FEMA assistance with floodproofing; calculator of household damage at various depths of flooding
	Elevate/floodproof home	Information on FEMA assistance, technical specifications, articulation of financial benefits, calculator of estimated damage with X feet of water
Individuals living in a "leveed area"	Develop emergency plan	Examples of emergency plans; height of potential flooding; evacuation routes; checklists for what to take and timeline
	Evacuate when requested	Marked evacuation routes, e-mail alerts, checklists for what to take, articulation of consequences of staying
	Observe levee for problems	"Levee Watch" program

(Continued)

Target Audience	Behavior Change Desired (examples only)	Information and Tools (examples only)
	Support Levee Safety Programs through resources (taxes) for operations and maintenance	Inspection reports, levee system assessments, stating consequences associated with deficiencies
Levee owner	Maintain reliable levees (e.g., O&M, rehabilitation) Inform public if levee is in danger of failing or overtopping	Inspection reports and assessments, make deficiencies public, better understanding of liability, state program enforcement
State and local governments	Develop and maintain robust levee safety programs	Information regarding number of people at risk, estimates of damage to critical infrastructure, economic impacts, need for compliance with National Levee Safety Program
Technical societies	Explain how levees are designed to work and limits of their use	Current standards and where problems with those standards are occurring; review of proposed new standards
	Advocate for funding required for levee infrastructure upgrades	Existing "lobbying" programs within societies; existing education and public awareness programs sponsored by societies
Developers, realtors, homebuilders	Promote floodproofing in new construction and renovation	Long term benefits to clients and customers and the sustainability of the community as whole
Media	Reporting on NLSP creation and progress Educate public about levee issues Develop a cadre of levee experts	Information about compliance, educate public about potential consequences of levee failure, statistics on what is protected by levees
School children	Increase geographical understanding of students protected by levees,	Education programs, field trips, incorporate into history and geography curriculum

Target Audience	Behavior Change Desired (examples only)	Information and Tools (examples only)
	awareness of benefits and risks, encourage parents to know how to evacuate and practice (similar to fire)	
Insurance	Provide financial breaks to those who take steps to mitigate damage through raising buildings, floodproofing, emergency plans	Mitigation measures that can be provided to customers

Figure 14. Example Target Audiences and Desired Behavior Changes.

Provide Technical Materials, Assistance and Training to States and Communities

Crucial to the successful adoption and consistent implementation of a *National Levee Safety Code* is a comprehensive and informative set of technical materials and direct technical assistance. This is particularly critical in the levee context since a majority of the levees in the country are outside the purview of the federal government. States and local agencies need to be provided the knowledge and the tools necessary to have an approvable levee safety program, particularly in the start-up phase.

The level of expertise with regard to the design, analysis and inspection of levees varies greatly across the country. The success of a national program depends upon common and highly sophisticated understanding of levee design and performance. The success of a NLSP is dependent on increasing the expertise and number of levee professionals across the country—hence a comprehensive training program.

The design, operation, and maintenance of levees are constantly evolving. With that evolution is the need to facilitate the flow of new and updated technical information. While conferences, technical assistance, and training are all proven methods to accomplish this, all three approaches in concert are more valuable.

The Corps is arguably the Nation's preeminent expert in levee design, analysis and inspection. A program that builds on that expertise (and lessons learned from the Dam Safety Program) will be the most effective and efficient. The Commission should work with the Corps to develop this three-pronged effort in developing and implementing: 1) technical materials; 2) training program; and 3) direct technical assistance. Specific recommendations can be found below.

Recommendation #10: The National Levee Safety Commission should contract with the Corps to take the lead responsibility and be provided the necessary funding to develop, maintain, and periodically update technical assistance materials dealing with state and national levee safety programs and the physical integrity of levees.

- The Corps has numerous technical publications that cover a broad array of technical information to include levees and related materials. The Corps should consolidate its published information pertaining to all aspect of levees (planning, design, construction, O&M, etc.) and make available on the NLSP web site and periodically update.
- The USBR should provide materials, expertise, and resources in developing technical assistance materials with respect to canal structures.
- The Corps' Engineering and Research Development Center (ERDC) should initially conduct a literature search for best practices pertaining to all aspects of levees and publish on the NLSP web site and periodically update. The materials should be organized in a manner that is easily accessible and usable. Over time, the responsibility for the content of the technical assistance materials should be led by the standing Technical Committee of the Commission.
- Advertise, promote and educate the public, state and local agencies, owners and operators on the material available, how to access, and how to utilize the information to establish a state levee safety program and address the physical integrity of levees.

This recommendation is dependent to some degree on developing the *National Levee Safety Code*. To begin to energize the states and for local governments and others to take a more active interest in levee safety, state and local entities have to be provided some tools with which to work.

Recommendation #11: Develop a national levee safety training program that includes the following minimum elements:

Flooding of Patrick Manor Senior Housing Community. Pocahontas, AR— Photo by Elmo Webb, PE 3/23/08.

- A specific curriculum, the successful completion of which would result in the certification of the graduate as a "Certified Levee Professional";
- Under contract with the Commission, the Corps should expand its current training program at either the Huntsville Center or Davis (HEC) to add classes in levee design, analysis and inspection. These classes should be made available to public and private sector. Consideration should also be given for the Corps to contract some of the training out to the private sector;
- National training opportunities— host recognized authorities in the engineering field to present and discuss analysis techniques, construction methods and other issues that can increase the expertise and information available to all engineers in the levee safety community;
- Local training through direct assistance to the states;
- Self-paced training; and
- Annual National Levee Safety Conference sponsored jointly by pertinent federal agencies (e.g. Corps, FEMA, and USBR) and/ or national flood management professional organizations (e.g. ASDSO, USSD, NAFSMA, ASFPM). This could be based on *The National Flood Risk Management: Levee Safety Summit* which was held in St. Louis, MO, in February 2008 (co-sponsored by the Corps and FEMA

and jointly hosted by ASFPM and NAFSMA)—a combination of information sharing and training opportunities.

Develop and Implement Measures and Practices to More Closely Harmonize Levee Safety Activities with Environmental Protection Requirements and Principles

For levees to perform adequately and reliably, it is essential to perform maintenance and rehabilitation activities before a project becomes functionally impaired or failures begin. Non-federal partners have had difficulties in the past obtaining the necessary permits to perform needed operations and maintenance activities on existing federally-partnered levees, many of which have operations and maintenance activities outlined in manuals developed and issued to sponsors before the passage of current environmental protection laws such as the Clean Water Act and the Endangered Species Acts. In order to better harmonize these perspectives and ensure that the protection of human life is not compromised, the Committee recommends a series of actions to better understand and remove barriers to effective levee operations and maintenance.

Recommendation #12: Develop and implement measures and practices to more closely harmonize levee safety activities with environmental protection requirements and principles.

- The Commission should direct Research and Development efforts to evaluate O&M practices for existing projects and to develop cost-effective measures to make O&M practices more compatible with present-day natural resource management principles. Development should be by an interdisciplinary team, comprising technical and environmental expertise, addressing the need to protect public safety and the need to protect natural resources.
- The Commission should establish a standing committee to address O&M for existing projects and to address how to better coordinate environment and safety issues on rehabilitation and new construction.
- The Commission should require states to establish an approach to facilitate operations and maintenance permits among each of the state resource agencies as part of a qualified program.

**Creating a Cadre of National Levee Experts:
Certified Levee Professionals**

In order to ensure a high level of professional training and experience and significantly expand the levee expertise needed to accomplish our national and local goals, delegation of the *National Levee Safety Program* (or parts thereof) to state and/or local entities should occur only if that entity has at least one "Certified Levee Professional" (CLP) on staff (or under contract) that is significantly responsible for the program. Such certification will only be granted to Licensed Professional Engineers with applicable expertise, experience, education, knowledge skill and ability in levee safety and who successfully complete this certification program. In addition, a provision for continuing education will be mandatory to maintain the certificate. Names and professional information regarding CLPs will be kept on file at the National Levee Safety Commission.

Conduct Research and Development to Support Efficient and More Cost Effective Levee Safety Programs

A major challenge facing those responsible for levees is conducting appropriate and rapid geotechnical assessments of levee integrity. These assessments are critical to providing assurances of levee safety. However, such assessments, depending on the nature of the material and the cross section of the levee, are commonly very costly. The bulk of the costs are related to the number and depth of soil borings. While some research is underway in Japan and the Netherlands on use of remote electro-magnetic sensors, no reliable methods or technologies are currently available in the United States to replace soil borings, with the principal exception being cone penetrometer soundings. Currently, very little effort is underway in the Research and Development (R&D) community to deal with this challenge. Early R&D efforts should focus on improvement of rapid assessment of levee geotechnical characteristics and integrity, and should consider research initiatives that would look at improved use of helicopter electromagnetic (HEM) and ground-based electrical resistivity surveys.

Conducting a dialogue with the most preeminent and influential members of the R&D community will bring together the best minds to help assure that an integrated, collaborative and comprehensive R&D program is developed and implemented. This will also provide potential sources of funding for the program.

There currently exists a large body of R&D knowledge both nationally and internationally that would be helpful to owners, operators, regulators, etc. Consolidating the body of knowledge and making the information easily accessible would be of great benefit and something that could be provided relatively early on. Assembling a working group to further develop a prioritized list of future R&D needs will help assure that the appropriate R&D is being conducted that meets the needs of all interested parties.

Recommendation #13: Develop a Research and Development (R&D) program funded at the federal level, and guided by a Standing Committee of the National Levee Safety Commission, that includes as a minimum:

- Innovative technology for repairs and improved engineering methods that would lead to more reliable levees and more cost-effective approaches.
- Technical and archival research— The Corps' ERDC should conduct a search of current technology for repairs and improved engineering methods, tools and products for dissemination. Assistance by the National Science Foundation to focus some of its research on improving rapid assessment of levee geotechnical performance.
- Dissemination of research products (e.g. technical manuals and guidelines, workshop and conference proceedings, training manuals, executive summary documents, brochures) to the levee safety community.
- Technology and tools to enhance the security of levees at the operation level.
- Establish guidelines and a program for the forensic investigations of levee failures and/or severe levee distress.
- A standing Technical Committee of the Commission should provide advice on program direction and priorities. The Committee should include representatives from academia, National Science Foundation, National Research Council, White House Office of Science and Technology, National Science and Technology Council, and the Corps' ERDC.

Design and Delegate Program Responsibilities to States

The foundation of a strong *National Levee Safety Program* is effective state and local programs. *As discussed in more detail in the next section, Building and Sustaining Strong Levee Safety Programs in All States,* states are best positioned to organize, implement and oversee levee safety programs within local communitees across the country.

They have a combination of necessary legal and taxing authorities, statewide reach and relationships to make programs successful. As with other national regulatory programs that require consistence and adherence to national standards (e.g., National Pollutant Discharge Elimination System, National Dam Safety Program) states need clear, rational standards, helpful guidance, training and implementation assistance, funding assistance and an open dialogue with the federal government. It is one of the most important roles of the Commission to develop an effective and efficient delegated program to states.

Major steps needed to develop and sustain a delegated program to states are:

Development of Standards

- Design and construction
- Rehabilitation
- Operations and maintenance

Susquehanna River floodwall being over-topped. Binghamton, NY. 2006—Courtesy of NYSDEC.

Development of Processes

- Inventory and inspection
- Risk assessment
- Improvements
- Oversight
- Enforcement

Program Elements

- Technical
- Legal
- Financial
- Administrative
- Institutional

Major Delegation Steps

- Develop guidance
- Provide technical assistance
- Communicate with and involve stakeholders
- Provide financial support/grants
- Review delegation plans and packages
- Negotiate
- Approve/disapprove
- Oversee
- Rescind state program (if necessary)
- Operate federal (regional) program for non-delegated states

Building and Sustaining Strong Levee Safety Programs in All States: The Cornerstone of a National Levee Safety Program

The National Levee Safety Act clearly indicates Congress' intent that state levee safety programs be created through delegation to better manage the critical life safety infrastructure associated with non-federal levees. The benefits of building and sustaining strong state levee safety programs are multiple:

- States are uniquely positioned to oversee, coordinate, and regulate local and regional levee systems as they already have such roles with regard to other elements of infrastructure and the environment. It is not appropriate or realistic to approach the management and oversight of local and regional levee systems from a single, national level. Allowing for a degree of variation and tailoring to meet local needs and circumstances rather than a national, one-size-fits-all approach is desirable.

- Coordinating and leveraging existing and complementary programs are already underway in some states.
- The authority for creating and implementing state levee safety programs rests with individual states, not the federal government.
- States are best suited to compel standards and good practices of local levee owners and operators.

Promoting Tribal Levee Safety Programs

Congress intended to include the participation of Indian Tribes in the development of a *National Levee Safety Program*. This is evidenced by the specification to have tribal representation on the National Committee on Levee Safety. Unfortunately, no tribal representatives were able to fully participate at the Committee level during the very short time period when the Committee was convened to develop this report. However, the Committee was able to benefit from review comments provided by tribal representatives. The Committee recognizes that tribes represent sovereign entities and that there are commonly many jurisdictional issues between tribes and other agencies. The Committee also recognizes that different tribes, as with different states, will have different capabilities in implementing levee safety programs. Nevertheless, it is essential that efforts be made to ensure that people living on tribal lands will also benefit from levee safety programs. The Committee believes that states and the National Levee Safety Commission will work collaboratively with tribes in developing levee safety programs, and that different approaches and arrangements will be developed on a case by case basis. The Committee looks forward to the participation of tribal interests in the further refinement of the recommendations encompassed in this report and in the development of a *National Levee Safety Program*.

Complimentary State and Federal Levee Safety Programs

In establishing and sustaining state levee safety programs, there are distinct roles for both the Commission (addressed earlier in recommendations) and for the states with delegated levee safety programs. States would operate such programs in conformance with national standards and requirements and provide timely and regular notification of their performance to the Commission. The Commission would, in turn, provide grants, training, technical assistance and guidance, clear national standards, and monitoring to ensure the success of the delegated programs. States with levee safety programs that exceed the minimum qualifications would receive additional incentives.

Physical Systems Approaches

Multi-jurisdictional programs are potentially a more effective basis for overall management of levee systems that do not lie entirely within any one political jurisdiction. States should be encouraged to cooperate with other state, local or federal entities to implement levee safety program elements for levee systems that cross jurisdictional boundaries. Such systems approaches are desirable because floods respond to physical systems—basins, protected areas, and major basin areas—not political or jurisdictional systems. The Commission would encourage systems approaches by providing additional incentives to states that implement NLSP elements through inter-jurisdictional cooperation agreements.

Principles of Delegation

States have primary authority for implementation of a levee safety program within their borders and it is the primary goal of the NLSP to delegate to and have strong state levee safety programs. The Committee recognizes that there likely will be instances where other approaches and delegation are necessary: (1) in the event that a state does not qualify for a delegated program, the Commission may consider designating local governments within the state to implement elements of the NLSP if the Commission judges such designation to be in the best interest of levee safety and/or conduct certain minimal levee safety activities via the Commission; (2) states may further delegate responsibilities for levee safety actions within their state; and (3) there are operations and maintenance requirements that belong at the owner/operator level and should not be assumed at the state or federal organizational level.

Key Elements of a State Levee Safety Program

Recommendation #14: Delegate implementation of National Levee Safety Program activities to qualified states.

Delegation should be highly encouraged, and therefore obtainable with qualifications necessary to perform the basic functions of the NLSP. The requirements of a State Levee Safety Program include three primary elements: legislating statutory authorities; implementing rules, regulations, and procedures; and securing resources for these activities.

Result of a levee break.Montegut, LA. 2002— Courtesy of FEMA.

Flood damaged levee. Bainbridge, NY— Courtesy of NYSDEC.

Susquehanna River at top of Floodwall. Binghamton, NY. 2006—Courtesy of NYSDEC.

Interior flooding and internal drainage. Endicott, NY. 2006—Courtesy of NYSDEC.

Mississippi River Levee. Midwest Flood. 1993— Courtesy of FEMA.

Statutory Authorities

1) To participate in the NLSP as established by the federal government.
2) To receive such funds as the federal government may make available for program implementation, and to distribute some portion of those funds to local government entities, consistent with the national program requirements.
3) To adopt or establish standards for levee classification, inspection, construction, operation, maintenance and emergency preparedness.
4) To perform or require performance of inspection of levees, and to prepare or require preparation and submittal of inspection reports and records.
5) To require or perform development and implementation of emergency action planning procedures.
6) To prepare reports of levees within the state, including location, condition, maintenance, areas protected, and risks posed thereby and to publish and distribute such reports to public or private entities.
7) To communicate with and educate local government and the public at large about the risks and benefits associated with levees and other flood-risk reduction measures, and to promote prudent practice with regard to levees.
8) To require that local government develop and implement emergency action planning procedures and evacuation plans for imminent or actual levee failure.
9) To enter public or private property for safety inspections or to perform emergency action.
10) To promulgate rules, regulations and procedures to implement these statutory authorities.

Rules, Regulations and Procedures

1. To coordinate levee safety activities among entities within the states owning, operating, regulating or using levees and between those entities and the NLSP.
2. To receive and review application packages from entities within the state for grants from the NLSP, to submit acceptable applications to the NLSP, and to receive and disburse grant funding from the NLSP.
3. To request an initial inspection by the Corps of the levees within the state's jurisdiction.
4. To inspect or require the inspection of the levees within the state's jurisdiction at least annually and after all significant high water

events. The inspections should be performed under the supervision of a registered engineer who possesses a levee training certificate from the NLSP.

5. To provide information to the national levee database for the levees within the state and to provide updates at least annually, following the standards for the database, including identifying the hazard potential classification of levees.
6. To implement a levee risk communication and public outreach /education program, including publication of an annual report on the State Levee Safety program, and on the results of levee inspections, and providing public notification of the maintenance ratings and risk behind levees.
7. Adopt the *Interim National Levee Engineering Guidelines*, and when available, the *National Levee Safety Code*, for all levee projects under state jurisdiction or involving state funds.
8. To require that all communities protected by *Significant* and *High Hazard Potential* levees develop emergency action and evacuation plans in accordance with NLSP guidance.
9. Adopt measures as needed to require consideration of nonstructural measures associated with any levee related activities.
10. To have a FEMA approved *Hazard Mitigation Plan*. Updates of plans should specifically reflect current condition and activities associated with levees.
11. To require that states provide liaison and coordination on environmental permitting actions.

Resources

Funding, qualified personnel, equipment and vehicles to conduct elements of a state program are the responsibility of states, local governments, and owners and would be principally provided by the states.

Absence of Delegation to Qualified State

In the absence of delegation to a qualified state program, the Commission should implement the following program measures:

- After an initial federal inspection and assessment, conduct or cause to be conducted an inspection of high or significant hazard levees after significant flood events, and at least every five years, and update the National Levee Database.

Dredge placing material for levee construction in California's Central Valley. circa 1920's—Courtesy of California Department of Water Resources.

- Provide inspection reports and findings to local emergency management officials.
- Conduct a program of public information concerning the presence of levees, their condition and their associated risks, including notification of the state legislature and governor.
- Other and further action as the Commission deems appropriate to encourage, publicize the benefits of and foster support for a qualified state program.

Philosophy of Incentives and Disincentives

The Committee recommends that the start-up period of the NLSP and delegated state programs be highly encouraged through both direct support (e.g. program start-up grants, technical assistance, training) with no penalties for non-participation. After the start-up period is complete and states have been afforded ample opportunity and assistance to ensure the safety of their populations through strong levee-related mitigation activities and the maintenance of reliable and resilient levees, an increasingly substantial set of disincentives should be applied.

Over time, increasingly stringent disincentives (e.g. lower priority for flood control funds) should be applied, making it more difficult for states and local governments to secure federal investment (e.g. public housing, schools) in areas located behind uncertain or unreliable levees. The Committee believes that this phased approach toward application of incentives and disincentives recognizes two strongly held and equally important beliefs:

- Significant time and assistance is needed for state/local governments and owner/operators to understand and address their levee situation (this problem took years to develop and will not be fixed quickly); and
- Continued federal investment in areas protected by levees that do not invest in protecting the people and property located behind them (e.g., participate in a minimum state program) is both fiscally irresponsible and places citizens at unacceptable risk.

National Levee Safety Program Requirements for Owners and Operators

It is the opinion of the Committee that it is most effective and efficient for owner/ operators to continue to be the primary responsible parties for crucial day-to-day activities. Recommendations to create a national program and delegated state programs do not take the place of the following key responsibilities of owner/operators (in some cases owners/operators are federal and state government agencies):

- Perform routine O&M including — routine inspection
  - routine maintenance
  - appurtenant works maintenance
- Perform on-site specific training
- Fulfill specific role in Floodplain Management Plans (in coordination with state and local governments)
- Local communication and education of risks
- Provide flood fighting and notification of distress
- Coordinate with local/regional flood fighting
- Participate in shared/new construction
- Perform repair, rehabilitation, replacement with sufficient property rights
- Develop and communicate emergency action plans (in coordination with state and local governments)

It is also the philosophy of this Committee that there are many additional levee related activities and responsibilities beyond minimum program requirements that should be performed at the state and local levels. Incentives should be offered to perform them. Because there is such a wide array of potential activities that may be utilized to increase the robustness of a state or local levee safety program, delegated programs that exceed the minimum requirements should be rewarded in proportion to the public safety benefits provided by the particular combination of activities they are performing. This could be addressed using a system of rewards like the Community Rating System, wherein a point-based system is applied to measure many different floodplain management activities and reward communities, through discounted insurance premiums, in proportion to the strength of the community's floodplain management program.

This document addresses incentives and disincentives in two main sections of this report. In the section *Financial Assistance Needed to Address Our Nation's Levee Problem*, the Committee describes two funding sources required to make state and local programs successful. The section entitled *Aligning Existing Federal Programs to Promote Effective Mitigation in Leveed Areas* recommends specific adjustments to three FEMA programs to limit federal financial exposure and reward good levee behavior. This section also suggests potential other areas of exploration as incentives/ disincentives for investing in levee safety programs.

Floodfighting and Internal Drainage. Oxford, NY. 2006—Courtesy of NYSDEC.

Please note that many of the incentives/benefits for state delegation as described in *Aligning Existing Federal Programs to Promote Effective Mitigation in Leveed Areas* can also be used as disincentives down the road. For example, eligibility and preference for P.L. 84-99 rehabilitation funds could be afforded to communities in states where there are qualified state levee safety programs. Conversely, lack of eligibility, lower priority or lower federal cost share should be afforded to projects in states that (at some point down the road) fail to create a qualified state levee safety program.

Financial Assistance Needed to Address Our Nation's Levee Problem

Considering the lack of understanding we have of the location and condition of our nation's levee infrastructure, the potential for catastrophic failure in some urban areas and the need for a coordinated, common approach to assessment, prioritization and risk reduction activities, the Committee proposes to Congress the need for two separate, but equally important sources of federal assistance. First, in order that the degree to which your levee safety is not dependant upon where you live, the Committee believes that federal funds should be expended to stand up levee programs in all 50 states with the degree of funding related to the hazard and complexity of levee safety in that entity. Second, the Committee proposes to Congress the development of an additional federal cost share program whose intent is to make more reliable and resilient existing levees as well as assess whether a structural solution is the most appropriate.

Commission Should Reward States and Local Communities Who Display Superior Performance

Experience with the NFIP and other federal programs suggest that states and communities benefit from availability of specific incentives to encourage best practices to exceed minimum program requirements. In the absence of such recognition, states and local governments operating within significant budget constraints often rely solely on minimum standards to comply with a federal program. Unfortunately, experience also teaches that reliance on minimum standards in the natural hazards risk management realm can have catastrophic results, such as to increase loss of life and property in disasters.

This recommendation would provide for incentives and disincentives for hazard reduction and mitigation.

These hazard reduction and mitigation activities can be far more effective at managing risk than rehabilitating or improving the levees themselves and are of enormous benefit both to the community and to the nation. Where feasible, they should be implemented as alternatives to levee work. Where levee work is occurring, these activities can be key elements of an overall flood risk management strategy for the leveed area.

State levee safety programs that exceed minimum requirements by permitting levee work and regulating new/enhanced levee construction will help to manage flood risk in the state and benefit the state and the Nation.

As part of the recommendation to support strong state programs, the Commission should identify, support, and incentivize best practices for states and communities to exceed minimum requirements for delegated levee safety programs and for managing risk in leveed areas. The National Levee Safety Commission, FEMA, the Corps, and other agencies should identify opportunities within their programs to reward states and communities for superior performance. A system of incremental rewards, through various incentives, should be developed to provide rewards to states and communities that are doing the most to manage their levee systems and their flood risk in leveed areas. The Community Rating System is a good example of such a system of incremental incentives/rewards linked to desired behavior or best practices.

States with successfully operating levee safety programs should be rewarded to the extent that their safety programs exceed minimum requirements, such as by:

- Requiring permitting or registration of all levee systems.
- Requiring compliance with the National Levee Code for all levee construction in the state.
- Requiring approval of design and construction of new levees and levee alterations.
- Performing levee construction inspections.
- Ordering procedural or operating changes, maintenance, repair, degrading, removal of encroachments, or removal of levees, where identified as the best measure for risk management.
- Performing or contracting for maintenance, repairs, emergency actions, degrading, removal of encroachments, or removal of levees.

- Taking over maintenance responsibilities of levees not being adequately maintained by a local owner/operator.
- Acquiring property rights (e.g., eminent domain) for levee safety, where necessary to prevent harm.
- Encouraging community participation in the NFIP and even exceedance of the minimum NFIP requirements (especially floodplain management behind levees).

States and communities should be rewarded when they exceed minimum requirements for managing flood risk in areas protected by levees. These include both nonstructural and structural alternatives, such as:

- Enhanced public involvement, outreach and notification regarding flood risk associated with levees.
- Enhanced involvement of levee owners/ operators to provide for opportunity for review, comment, and approval of proposed development behind the levee.
- Notification to prospective buyers in leveed areas of flood risk behind levees, state's status in the NLSP, and community's status in the NFIP impacting availability of federal flood insurance.
- Public notice state's status in NLSP.
- Public notice of community's status in NFIP and availability of federal flood insurance.
- Promotion or requirement of flood insurance purchase.
- Contribution of locally generated data regarding levees to floodplain mapping.
- Levee hazard mitigation activities as part of an enhanced community or state levee safety or hazard management plan, which may include:

  - Buyouts/relocation of structures
  - Elevation of buildings
  - Floodproofing of structures
  - Enhanced building codes
  - Enhanced land use, zoning, and local community planning to prevent intensification of development behind
  - levees contrary to tolerable risk
  - Preservation of open space to allow for flooding, and to prevent

harm in the event of levee overtopping or failure
- Requirement of flood water retention/detention areas, constructed wetlands, and similar nonstructural flood risk reduction measures

- Reservoir reoperation.
- Channel enlargement.
- Require community participation in the NFIP.

Grants to Create Levee Safety Programs in All States

Many states and communities have difficulty raising funds for levee safety activities. Levee safety program activities that assist individuals and local governments in better measuring and understanding risk to human health and safety and better cost estimates of potential flooding damages will make this easier over time. However, in the interim, to make the NLSP achievable, states will need funding to get the program up and running and to keep it sustainable.

The consideration for grant prioritization for *National Levee Rehabilitation, Improvement, and Flood Mitigation Fund* described in the next section will provide a great deal of incentive for most states and local governments. This, in combination with these start-up grants, will likely incentivize states to implement a levee safety program sooner.

Upholding the adage that an *ounce of prevention is worth a pound of cure*, the Committee believes that federal investment in setting up state safety programs will reap a many fold reduction in the need for federal disaster assistance, reducing the overall federal investment.

Recommendation #15: Establish a new levee safety grant program to assist states, local governments and owners and operators to achieve strong levee safety programs.

The Committee envisions that state levee safety programs will include state and local levels of government working cooperatively to accomplish the program goals, with a division of responsibilities as each state and local government is able to decide. Federal funds to assist state levee safety programs would therefore flow to the agency that is actually performing the federally funded work. It is intended that much of the funding would be delivered through state programs to the responsible agency performing functions such as inspections, preparation of reports and emergency action planning.

Thus, it is envisioned that much of these funds would end up flowing to levee owners/operators and to the local agencies. Consequently, one of the requirements for a delegated state program is the ability to manage and disburse federal grant funds. Further, the administration of grants by the Commission to carry out this work must help verify that grant funds are used to reimburse the actual agencies that are completing the tasks associated with state levee safety programs.

- Note: While the Commission is being created, FEMA should administer the grant programs on their behalf. Once the necessary processes and resources are in place to properly administer this activity, the Commission would assume responsibility.
- Additional support/funding could be provided to states to support to multi-jurisdictional or levee system-specific programs.

Raising Funds to Support Strong Levee Safety Programs

While federal grants will be critical for establishing and maintaining strong levee safety programs within states and local communities, it will continue to be necessary for states, communities and levee owners to raise funds to conduct necessary state, local, and owner/ operator activities in perpetuity. The people that live, work, and own property in leveed areas are the most direct beneficiaries of levee program safety activities and should be the primary source of funds for upkeep and mitigation activities. Further, funds generated at the state/local level are critical for healthy safety programs and can often serve as the nonfederal match for federal cost sharing opportunities. The examples below describe two existing state approaches to funding levee safety activities:

State of Texas: Texas State statute provides for collection of fees on flood insurance premiums, generating $6.2 million biannually to support floodplain management throughout the state.

State of California: The State of California passed two major bond initiatives in 2006, authorizing $4.9 billion for flood management activities. Most of the bond funding is for repair and improvement of levees, with requirements for local cost sharing to match the state funds. Approximately $15 million per year supports maintenance of certain levees in the Sacramento-San Joaquin Delta.

Like the Texas approach, states could require a fee on flood insurance premiums sold in leveed areas (e.g., AL and XL zones) and use the generated funds for the levee safety program.

Caution would need to be exercised in establishing such fees in areas behind accredited levees (e.g., XL zones) prior to implementation of mandatory flood insurance in these areas, because doing so may reduce the number of voluntarily purchased flood insurance policies.

National Levee Rehabilitation, Improvement, and Flood Mitigation Fund

The *National Levee Safety Program* legislation being proposed will help enhance public safety by:

- Creating a National Inventory of Levees with Inspection Information
- Establishing National Levee Safety Standards
- Requiring Levee Safety Programs in All States
- Requiring Inspections and
- Assessments of Levees
- Funding Research to Enhance Technical Expertise for Levees
- Establishing Training Programs for Levee Safety
- Educating the Public, Levee Owners and Others About Risk and the Need for Strong Levee Safety Programs.

While the NLSP will contribute to reducing the risk to life and property and help improve the safety of our nation's levees, the safety of levees demands much more attention from national policymakers. This program basically establishes only the minimum effective management program for the nation's levees and related infrastructure. By itself, the NLSP does not provide funding to address the many levee deficiencies that are expected to be discovered and documented.

Failures and devastation will continue to occur and threaten this nation as levees continue to age and deteriorate and as urban populations grow and development behind levees increases. Because of increasing population and development behind levees, the risks are expected to actually increase over time even if modest levee improvements are made. Failures affect large populations, flood into neighboring states and cost millions of dollars in federal disaster relief spending. There are likely many thousands of miles of unreliable levees throughout the United States. Events over the past two years illustrate the catastrophic results that can occur. The eyes of the nation were focused on the catastrophic consequences of Hurricane Katrina in New Orleans.

The management processes contained in levee safety programs, in and of themselves, do not solve problems that continue to grow as levees deteriorate and needed rehabilitation to bring them up to current safety standards is deferred. The priority on rehabilitating our aging and deteriorating national infrastructure must include levees. In 2006, the State of California passed two bond measures that would provide $4.9 billion for levee and other flood protection repairs and improvements.

However, this figure pales in comparison with the $30 billion experts say would be needed across the state. A review by Scripps Howard News Service of levee oversight and funding at the state and national level suggests the new focus still may not be sufficient to overcome decades of neglect.

The expansion of a *National Levee Inventory* will further enhance the recognition and realization of the deteriorating condition of many of the nation's levee structures and of the lack of a focused public policy to address the problem. Federal, state and local levee owners will then need a funding source to assist with rehabilitating our aging and deteriorating levee infrastructure and correcting decades of neglect. It is difficult for many levee owners to find the funding necessary to undertake rehabilitation work when necessary. Often, vital repairs are neglected, and these levees are subject to further deterioration due to lack of funds and neglect. Deterioration can lead to levee failure. These types of disasters can cause great destruction and loss of life, with no respect for state boundaries. A few states across the country, such as the State of California, have established innovative funding programs but there is currently no comprehensive federal funding mechanism to assist levee owners. Levee districts, like many levee owners, are strapped for cash, especially the large sums needed to finance costly levee repairs. The challenge at federal, state and local levels continues to be securing adequate funding countrywide for levee rehabilitation.

Dallas Flood Control protecting downtown Dallas. Dallas, TX—Courtesy of City of Dallas Flood Control District.

Key questions before the American people are:

- Will the federal government find a way to assist levee owners or will future catastrophic levee failures with resulting property damage and loss of life continue to occur?
- Will the nation learn from the experience of Katrina that it is far better to invest in levee rehabilitation rather than disaster relief and recovery? (i.e. pay me now or pay me more later)

It is a reasonable expectation of every US resident to be protected from preventable disasters such as levee failures. There is a critical need to create a federally administered levee rehabilitation and flood mitigation program in order to repair our nation's unsafe levees. Additionally, paralleling such a federal initiative should be similar efforts for state and local governments to create their own loan or grant programs for levee rehabilitation. There is a great need to begin an assistance program at both federal and state levels to help levee owners with their rehabilitation needs. This is a public safety issue.

Recommendation #16: Authorize the National Levee Rehabilitation, Improvement, and Flood Mitigation Fund A federally authorized program should be developed and cost-shared (65% federal and 35% state/local) for non-federal publicly-owned levees.

Funds would be available to address both structural and non-structural measures so long as the combination of measures maximizes overall risk reduction. Provisions could be made where a percentage of the nonfederal cost share could be met through implementation of nonstructural measures. This program would only be authorized for pre-disaster declaration and would not replace or substitute FEMA Mitigation Program funding. The legislation would provide funds directly to states based on a screening level risk-informed priority system that would be based in part on information taken from the NLD. Such federal assistance would initially be limited to only levee systems that protect existing urban areas that have a high damage potential.

Eligibility for this funding would have several requirements to assure that owners/operators maintain a high level of upkeep of their levees and engage in responsible activities related to the public protected by those levees. In order to be eligible to receive federal assistance a grant applicant must:

- Provide the minimum data to populate the National Levee Database;
- Demonstrate the financialmeans to provide their cost share contribution for the initial rehabilitation and the financial assistance to operate and maintain the levee system in accordance with the *National Levee Safety Code*;
- Evaluate an array of non-structural alternatives/activities, and where applicable identify a nonstructural/ structural blend of flood risk management approaches, and dmonstrate that the appropriate combination of measures are being implemented to best reduce flood risk;
- Engage in public outreach/ notification;
- Provide buyer notification of flood risk;
- Promote purchase of flood insurance;
- Develop an emergency response plan;
- Develop and implement an *Inspection of Completed Works* program;
- Provide a flood risk management plan as part of a public safety element of a general/master land use plan that demonstrates the local community plan to manage land use over time to move substantially towards the established national tolerable risk guidelines; and
- Participate in the NFIP or be located entirely within one or more participating communities.

Although the 1%-annual-chance (100-year) flood insurance standard required by the NFIP does not embody a levee safety standard for protection of life and property, participation in the NFIP demonstrates the community's commitment to review development and enforce at least the minimum standards of the NFIP to prevent harm in and around its floodplains, including areas of residual risk associated with levees.

The federally sponsored levee safety program would be established through legislation that would be enacted at the same time as the Commission. Early funding could be used to assist states and local interests in conducting levee evaluations that will help inform the condition of levee systems and further facilitate funding priorities. It is anticipated that it will take two years for states to populate the NLD and develop a risk-based tool that would be used to assist in prioritizing the allocation of funds. The authoring language would, at a minimum, spell out the 65/35 cost-sharing provision; minimum requirements for a state to be eligible for assistance; and further specify that Congress rely on the recommendations of the Commission on the priority of allocation of funds based on the NLD and risk-based assessment performed and the level of appropriations over the next five years.

Aligning Existing Federal Programs to Promote Effective Mitigation in Leveed Areas (Incentives and Disincentives)

All Federal Agencies Should Adopt the Letter and the Spirit of National Levee Safety Program

First and foremost, all federal agencies should adopt the *National Levee Safety Code* and comply with all other requirements of the NLSP for levees under their jurisdictional control. Federal agencies with expertise may be called upon to provide technical or programmatic guidance, assistance, support, and applicable training in the development and implementation of the NLSP. Federal agency adherence to NLSP requirements is important in that it promotes nationwide consistency in important technical standards, common approaches and messages related to risk communication/public education and improved coordination and harmonization of federal levee-related programs and requirements. Except for a few cases where new authorities might be called for, federal agencies could use their existing authorities to perform these activities.

Aligning Existing Programs

As mentioned in the previous section, *Financial Assistance Needed to Address Our Nation's Levee Problem*, grants should be provided to encourage states to support the set-up and maintenance of levee safety programs and to perform basic activities such as: update and maintenance of basic inventory, inspection, reporting, notification/ public outreach, and coordination.

Additional support should be provided for the costlier task of rehabilitating and improving levees, as well as the critical assessment of whether levees are the best flood risk mitigation option in a given situation.

In order to ensure that these investments have the greatest possible impact, all federal programs that significantly impact governmental and individual decision-making in leveed areas must be aligned toward the goal of reliable levees, an informed, involved public and shared responsibility for protection of human life and mitigation of public and private economic damages. Federal programs should not only be aligned with each other, but can be used as an enticement (benefits to be accrued upon the development of a state safety program) to responsible levee stewardship. Alignment incentives fall in the following broad categories:

1. Savings/funding to community
2. Eligibility for federal funding
3. Priority for federal funding
4. Cost sharing requirements

In addition to the two funding programs mentioned in the previous section, and three specific FEMA alignment recommendations detailed in this section the Committee recommends the following action.

Levee damaged due to overtopping. Hurricane Katrina. St. Bernard Parish, LA. August 2005.

Recommendation #17: Existing federal programs should be considered for use as possible additional incentives or disincentives to governments and their citizenry that have delegated state levee safety programs, per the requirements set forth by the Commission. For most of the examples below, incentives or the inverse (disincentives) can take the form of the four broad categories noted above (e.g., savings, eligibility, priority or cost share). Benefits from any given incentive may accrue at numerous levels, but it is possible to identify the targeted beneficiaries of the identified potential incentives, as shown below.

The Committee developed the existing recommendations under consideration of the following principles:

- Immediate disaster response functions should not be included as incentives and disincentives. Namely FEMA's Public Assistance Program Categories A and B and the Corps Flood Fighting function under P.L. 84-99 should be available to all communities in the face of a natural disaster. To withhold such immediate funds is inhumane, flies in the face of public safety, and does little to promote levee safety behavior.
- Ensuring that promoting synergies between the National Levee Safety Program and the NFIP do not result in unintended consequences. Links that are too strong between NLSP and the NFIP may further solidify the dangerous untrue belief by some that the 1%-annual-chance event (100-year) is a "safety standard" (see page 10 for a more in-depth discussion of this challenge). Further, any recommendations that include the NFIP must consider how all program components (hazard identification, insurance, and other mitigation actions) will work together. If they are not considered together there may be serious unintended consequences.

There are three specific recommendations related to the alignment of federal programs: 1) require risk-based flood insurance in leveed areas; 2) enhance FEMA's mapping program to communicate levee risk; and 3) align FEMA's Community Rating System (CRS) to reward good levee safety behavior.

Incentives/Disincentives	Property Owners	Levee Owners and	Local/ Regional	States
FEMA Disaster Assistance (non-emergency) Funds (e.g. Individual Assistance, Public Assistance Sections C-G, Mitigation Grants)	X	X	X	X
Corps P.L. 84-99 Rehabilitation Projects		X	X	X
Federal funds for infrastructure behind levees (e.g. Highway Funds, HUD grants)			X	X
Small Business Administration loans for disaster recovery behind levees	X			
Federal Loan Guarantees for disaster recovery behind levees	X	X	X	X
Federal flood controls projects from the Corps (General Investigations for new authorities & Section 216 for continuing authorities)		X	X	X

Note: Proposals to make changes in existing programs are intended to be revenue neutral. In the absence of new requirements, the intent of the Committee is that the funding for programs in this table remain largely the same, but that distribution of funds, preferences, etc. change as a result of beneficial levee safety practices. This approach is fiscally responsible in that it increases federal investment in communities whose levee safety programs (e.g., evacuation, land use, insurance) are more protective of human health and safety. Conversely, it reduces investment in the communities who forgo good levee safety practices.

Figure 15. Exploring the Need for Potential Incentives and Disincentives Through Existing Federal Programs.

Mandatory Risk-Based Flood Insurance in Leveed Areas

Flood insurance is one of the most effective ways to limit financial damages in the case of flooding and speed recovery of flood damaged communities. Currently, many people who live in leveed areas do not believe they need flood insurance as they are protected by a levee structure. This recommendation aims at increasing the understanding that living behind even well-engineered levees have some risk (sometimes referred to as residual risk). Implementing this recommendation will result in a greater number of home and business owners being protected from catastrophic financial loss. Further, this recommendation will increase risk awareness and preparedness of the public residing behind well-engineered levees.

The Committee believes that implementing this recommendation will incentivize communities to exceed the 1%-annual-chance (100- year) protection standard that has mistakenly become a target minimum. Because premiums would be risk-based, greater protection, through better, more reliable levees or better floodproofing programs would result in more favorable premiums. A similar proposal is contained in legislation proposed in Congress (H.R. 3121, Section 107. Mandatory Coverage Areas) and is supported by this Committee.

Recommendation #18: Require phasing in mandatory purchase of flood insurance for structures in areas protected by levees with risk based premiums.

Legislation would be needed to authorize mapping of residual risk areas behind levees and to enact mandatory purchase requirements in these areas.

- FEMA would be required to develop appropriate risk-based premiums.
- FEMA would likely publish revisions to the FEMA Mapping Programs requirements and NFIP regulations on a schedule that may be set by Congress.

Please note: Due to the differences in potential failure consequences, function and ownership, the Committee recommends that mandatory flood insurance not be required behind canal structures that do not have a significant role in providing hurricane, storm, or flood protection.

Enhance FEMA Mapping Program to Communicate Levee Risk to Communities

Identification of levee system consequence zones associated with levee failure will aid in determining hazard classifications, properties targeted for public outreach, funding, evacuation planning, mitigation, and other program components. The zones will set the boundaries for application of the NLSP.

FEMA is well-positioned to assist in levee risk communications because the NFIP flood maps (FIRMs/DFIRMs) are a primary source that local/regional/state entities access to assist in making local land use decisions.

The likelihood of a community implementing requirements associated with additional FEMA data is increased by use and access to FIRM/DFIRM maps. These maps consolidate much of the information into the place where decision makers already go to find related data. FEMA's website and resources are also frequently accessed by state professionals, mortgage lenders, prospective buyers, and property owners in reviewing property purchases.

Recommendation #19: FEMA's flood hazard mapping program should be augmented to include the following activities to further support National Levee Safety Program activities, especially those associated with risk identification and communication in levee system impacted areas.

- Identify levee systems, including structures along canals, and associated levee system failure consequence zones. This should be carried out in accordance with the development of the NLD, which will provide additional information on consequence areas behind levees. The completion of this step is dependent on and should be informed by the recommended inventory and inspection of nonfederal levees.
- Re-designate on DFIRMs existing Zone A/AE or Zone X areas impacted by levees as either AL or XL, respectively, to better communicate the greater flood risks in levee system impacted areas.
- Depict on FEMA's website additional flood hazard information (e.g. 200- year and 500-year floodplain maps) that may be provided by local/ regional/state entities.

Align FEMA's Community Rating System (CRS) to Reward Development of State Levee Safety Programs

The intent of FEMA's Community Rating System (CRS) is to reward communities that do more than meet the minimum NFIP requirements to help their citizens prevent or reduce flood losses. Through CRS Activity 620, the CRS also provides an incentive for communities to initiate new flood protection activities. By increasing the credit for levee safety activities, this recommendation would provide additional incentives to operate compliant levee safety programs. It would also reduce flood insurance premiums as they are based on risk, providing benefits directly to property owners and throughout participating communities and more importantly, reduce the overall hazard/damage potential. In order for this recommendation to be most effective, FEMA may also have to make the application process more user friendly and consider removing the construction date requirement.

Recommendation #20: The National Flood Insurance Program (NFIP) Community Rating System (CRS) Program should be revised to credit a community based on its state levee safety program and augmented to increase/decrease maximum credits allowed for certain CRS activities, including but not limited to Activity 620. The NFIP CRS Taskforce should revise CRS Activity 620 "Levee Safety" to:

- Provide credit for any community or communities within a state with a nationally compliant state levee safety program that has submitted the necessary documentation of its program to FEMA.
- Eliminate the requirement that CRS credit can only be provided to levees built before January 1, 1991.
- Eliminate the requirement that CRS credit can only be provided to levees that provide protection between the 4%-1%-annual-chance flood elevation.
- Increase the overall maximum allowable CRS credit that can be provided to any community for this activity, specifically for the operation, maintenance, and emergency/evacuation plan elements.
- Provide CRS credit to a community or communities within a state if the local/state hazard mitigation plan includes a list of all high hazard levees in the community/state and mitigation measures for the hazards they pose to the community or state.
- Revise method for calculating each of the elements of Activity 620.

The CRS Taskforce should consider revisions to other CRS activities as necessary to provide credit for certain levee safety program activities/elements, such as:

- Series 300—Public Information ◦ 330—Outreach Projects
    - 340—Hazard Disclosure
    - 350—Flood Protection Information
    - 360—Flood Protection Assistance
- Series 400—Higher Regulatory Standards
- Series 600—Flood Preparedness
    - 610—Flood Warning Program

FEMA Digital Flood Insurance Rate Map.

INVESTING IN A NATIONAL LEVEE SAFETY PROGRAM

Although the National Levee Safety Program comes with a cost, the overall proposition is: "Pay me now, or pay me even more later"

Donald Basham,
Committee Member

"and later could be tomorrow"
Craig Kennedy,
Committee Member

INTRODUCTION

One of the dichotomies of levees is that, while these structures have afforded the country economic prosperity, they have also had the unintended consequence of obligating the US taxpayer to pay disaster damages and repairs when these same levees fail. The average yearly national cost can run in the billions as evidenced by the recovery efforts still underway today in New Orleans. The potential risk exposure in the future is even greater. A National Levee Safety Program (NLSP) is not just a cost; it is a long term investment in public safety and continued economic prosperity. With growing development and consequences in almost all areas behind levees, the benefits of a strong safety program will only increase.

Public Safety

The primary benefit of a NLSP is the protection of public health and safety. Some would argue that the protection of human life is fundamentally an economic issue while others would suggest that you can not put a value on human life and the loss of even one life is unacceptable. Hurricane Katrina and the estimated 1,800 fatalities associated with both the storm and the levee failures is the best and most compelling example in support of a NLSP.

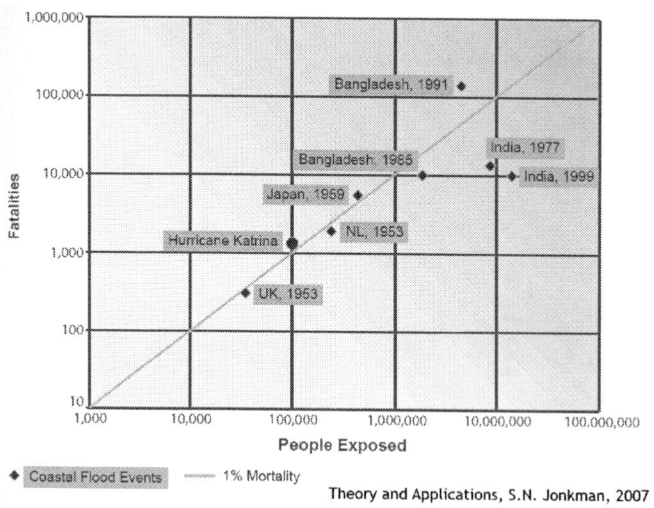

Figure 16. Loss of Life Estimation in Flood Risk Assessment.

As the events in New Orleans bore true, fatality rates for major urban areas due to flooding have historically been in the 1% range worldwide.

The exposure in some of the larger cities of the United States has the potential to match or exceed the catastrophic loss of life experienced in the Gulf Coast area in 2005. The very large events that would cause this type of loss of life have yet to be fully experienced in the United States simply because we have such an abbreviated history in comparison to some of the international communities. The table below shows lives lost due to major flood events that included levee failures in this country.

At the individual level, a robust levee safety program will not only inform people living behind levees of their risks but will engage and involve them in the process of risk communication, education, and awareness. An informed and involved public can participate in the shared responsibilities of risk management at both the individual level and the community level. Recent examples demonstrating the benefits of an informed and involved public include the effective evacuations of more than 2 million people from the greater New Orleans area in advance of Hurricane Gustav, and nearly 1.1 million people from the Texas Gulf Coast ahead of Hurricane Ike, both during the 2008 hurricane season.

Failure	Loss of Life*
Okeechobee Hurricane, September 1928	2,500
The Great Flood, 1929	246
Vanport, Oregon, 1948	16
Kansas-Missouri Floods, 1951	28
Yuba City, Yuba County, California, 1955	38
Northern CA & Northwestern Nevada, 1986	13
The Great Flood, 1993	47
Arboga, Yuba County, California, 1997	3
Hurricane Katrina, 2005	1,810
Midwest Flood, 2008	24

* Not known to be attributable entirely to levee failures.

Figure 17. Major Flood Events That Included Levee Failures and/or Loss of Life.

Financial Exposure in Leveed Areas —Bracketing the Cost

While preservation of human life is the most compelling reason for levee safety, a responsible public must also consider the benefits and costs of the NLSP. One of the challenges in trying to quantify the nation's flood risk with respect to property damage and economic loss is the lack of comprehensive information, particularly given the unknown number of levees across the nation and the unknown risks associated with them. Much of the available information on past flood damage and economic loss has been only partially captured, is often tracked differently by different agencies, and does not distinguish between flood damages in leveed areas and non-leveed areas. Nevertheless, some insight can be obtained by reviewing some of the available flood damage information associated with recent flood disaster events.

Corps Data

The Corps has compiled flood damage data associated with federal flood control facilities between 1998 and 2007 (Annual Flood Damage Reduction Report, provided by CECW-CE, 2007). During this ten-year period, flood damages associated with federal flood control facilities averaged $4.2 billion per year, excluding those associated with Hurricanes Katrina and Rita. Based on current information, it is reasonable to assume that about half of this was related to the 14,000 miles of federal levees, or about $2.1 billion per year. If this amount was then extrapolated to the estimated 100,000 miles of non-federal levees in the nation, the annual expected damage would be approximately $15 billion per year. However, Corps levees generally protect areas of more concentrated population, commerce, and infrastructure than the average non-federal levee. On the other hand, this compilation excluded the costs associated with Hurricanes Katrina and Rita. A reasonable upper bound limit for expected damage may be on the order of $10 billion per year.

State of California Data

The State of California has compiled flood damage data associated with state-federal project levees in California's Central Valley for flood events between 1955 and 2005 (California Department of Water Resources, Division of Flood Management). The average flood damage associated with these 1,600 miles of levees was found to be $70 million per year (in 2005 dollars).

If this amount was extrapolated to the estimated 100,000 miles of non-federal levees in the nation, the annual expected damage would be approximately $4.4 billion per year. While these Central Valley levees are typically major levees, the flooding was generally associated with agricultural areas and/or small rural communities. No major urban flooding was associated with these events. So, perhaps this extrapolation might be on the low side. A reasonable lower bound limit for expected damage may be on the order of $5 billion per year.

It is recognized that the above examples and extrapolations are not comprehensive and that they employ only simple calculations that do not tell the whole story. Nevertheless, they indicate that the annual financial loss associated with the nation's levees may be on the order of roughly $5 to $10 billion per year.

Insurance as a Basis for Exposure

While this Committee believes that a national levee safety program is a necessary investment and will provide significant reductions in the nation's flood risk behind levees, flood insurance will remain the most certain individual economic risk mitigation/reduction avenue available to citizens living and working within leveed areas (data shows that individuals with flood insurance are more easily and quickly able to recover from the devastating financial effects of flood disasters). Insurance data can also be used as a basis to roughly estimate the national financial exposure due to flooding.

Less than 6 million people currently hold flood insurance policies in more than 20,000+ communities across the United States. More importantly, it is estimated that only 10% of structures behind levees have flood insurance, and of those, most are not covered to the complete value of the property (both structure and contents). This demonstrates that the remaining 90% of the structures behind levees without insurance represent a significant exposure to the federal government in potential disaster assistance and recovery cost. Based on best available data, the current value of residential and commercial properties (structures and contents) located in all leveed areas alone constitute a total national cost exposure of more than $375 billion. An annual loss of $5 to $10 billion corresponds to about 11/2% to 3% of the total exposure.

Flood damaged levee. Bainbridge, NY—Courtesy of NYSDEC.

Losses Incurred from Past Events

Another means available for understanding costs both in terms of human life and dollars is to examine the data available from past documented flood disasters. The following synopses highlight some of those events.

The Great Flood of 1993

During the spring and summer (April— September) of 1993, extremely high rainfall occurred on the upper Mississippi River Basin causing major and/or record flooding for nine states in the upper Midwest. This event came to be known as "The Great Flood of 1993." The magnitude, severity, and longevity of this flood were extreme. It was wide spread, covering nine states and 400,000 square miles. Also, the flood was of extremely long duration, lasting nearly 200 days at some locations. In terms of rainfall amounts, record river stages, extent of flooding, persons displaced, crop and property damage, and flood duration, the Great Flood was the worst hydro-meteorological event to occur since the United States started to document weather events in the late 1800s.

Damage caused by these record flood stages was massive. More than 200 counties were declared federal disaster areas, including all 99 counties in Iowa. More than 31,000 square miles of land were inundated by flood waters. An estimated 72,000 private homes were washed away or suffered major damage. Between 35,000 and 45,000 commercial structures were damaged.

Along the length of the Mississippi River that forms the western boundary of Illinois, more than 1,000 miles of roads were closed and nine of the 25 non-railroad bridges were shut down and 12 commercial airports were closed by the flood. Additionally, the Corps reported that 40 of 229 federal levees and 1,043 of 1,347 non-federal levees were overtopped or damaged during the flood. There were also 15 flash floods triggered from these storms that caused dam breaks, the majority of which were in Wisconsin. Even in light of this, federal flood control efforts in the Mississippi basin prevented nearly $20 billion in potential damages. Estimates set the losses from this flood at $15.6 billion (1994 dollars) and this cost does not include all of the economic losses or the non-quantifiable, human impacts of this disaster.

Agriculture accounted for over half of these damages. Flood response and recovery operations cost more than $6 billion. Also, because flood insurance was not extensively used, it was estimated that 15% to 25% of the flood disaster costs were borne by state and local governments, not to mention the costs to uninsured homeowners who were forced to rebuild using their own resources. This natural disaster killed 47 people and forced 74,000 people from their homes.

Hurricanes Katrina and Rita, 2005

Hurricanes Katrina and Rita devastated the New Orleans area and wrought approximately $200 billion in damage and economic losses. Prior to these hurricanes, different parts of New Orleans probably had different levels of flood protection. However, for discussion purposes, the overall level of flood protection was probably on the order of about a 2%-annualchance, or about a 50-year level of flood protection. At face value, this could be interpreted to mean that the New Orleans area would have had an annual damage exposure of about $4 billion per year prior to Hurricane Katrina. However, this is too high since Katrina was a larger storm than a 50-year event. So, for discussion purposes, let us assume that the pre-Katrina annual damage exposure was on the order of $1 to $2 billion per year. Following these two hurricanes and the resulting devastation, the Corps is in the process of spending approximately $15 billion to repair and improve the area's levees and floodwalls. This investment is expected to lead to a 1%-annualchance (100-year) rated level of flood protection, and a 0.2%-annualchance (500-year) level of flood resiliency (i.e. floodwalls and levees expected to remain intact even if overtopped to this level of flooding).

Using the same set of consequences, this higher level of flood protection would roughly correspond to about a $400 million per year annual damage exposure—a significant reduction in future costs for this major urban area. The lessons from these events include:

- The roughly estimated $1 to $2 billion per year annual damage exposure prior to Hurricane Katrina is a tremendous exposure, and was only for one metropolitan area. There may be other metropolitan areas that have exposures on the same order of magnitude.
- The $15 billion being expended by the Corps to upgrade the flood protection system is a wise investment that will be repaid many times in avoided costs.
- Even after this investment and improvement in flood protection, there will remain a significant annual damage exposure of approximately $400 million per year. Again, this is still a relatively high number for just one metropolitan area and further supports the rough estimate of $5 to $10 billion per year for the nation as a whole.

Midwest Flood 2008

Midwesterners who experienced the Great Flood of 1993—estimated to have been a 500-year flood at the time—may have believed that they would not see another flood of that magnitude in their lifetimes.

Following the devastating hurricanes along the Gulf of Mexico in 2005, most Americans probably believed the country to be “in the clear” from flooding for at least a few years, if not longer, but unfortunately that assumption did not hold true.

During the summer of 2008, the Midwest once again experienced significant flooding following months of heavy precipitation. A number of rivers overflowed their banks for several weeks at a time and broke through levees at numerous locations. States affected by the flooding included Illinois, Indiana, Iowa, Michigan, Minnesota, Missouri, and Wisconsin. Approximately 35,000— 40,000 people were evacuated from homes. Flood warnings covered a span of about 325 miles from Dubuque, Iowa to St. Louis, Missouri. The Mississippi River crested at 37 feet in the St. Louis area, seven feet above flood stage.

Flooding continued for as long as two weeks with central Iowa, Cedar Rapids being hardest hit. In Iowa alone, nine rivers crested at record levels, 83 of 99 counties were declared disaster areas, and Iowa’s agricultural economic losses are estimated to exceed $2 billion. In Cedar Rapids, Iowa, flood waters covered 1,300 city blocks, inundating city hall, the county jail, the fire department, police communication equipment, most of the public library’s collection, and 3,900 homes. The Cedar River flood crested at over 32 feet, exceeding the historic 1929 record, and nearly six feet above the so-called 500-year flood level. Only 777 of the 4,000 homes damaged or destroyed by flooding were covered by any flood insurance.

The flood left two dozen people dead and damage region-wide was estimated to be in the tens of billions of dollars. To date, $2.7 billion in federal flood relief has been approved, but does not include the federal investment of low-interest loans or the value of crop insurance and private insurance payouts.

The above examples do not provide data for hard analysis of annualized loss of life or rate-of-return on levee project investments; however, they do underscore recent examples of the type of events that support the bracketed estimate of $5—$10 billion dollars per year in flood damages.

Chenango River. Near top of the floodwall during flood of record. Binghamton, NY. 2006—Courtesy of NYSDEC.

Need for Future Data Coordination and Management and Analyses

As noted previously, because there is great uncertainty in the scope of the national levee portfolio, there can only be marginal confidence in an estimate of costs associated with this portfolio until such time as a comprehensive inventory and assessment of levees is completed. The Committee fully believes that a comprehensive national inventory can be used to enable the development of a more detailed estimate of how much annual savings could be realized through the implementation of a national levee safety program. As stated above, much of the available information on past flood damage and economic loss information has been only partially captured, is often spread out and tracked differently by various agencies, and does not distinguish between flood damages in leveed areas and unleveed areas.

The Committee spent significant time collecting and examining various available data in its existing formats relative to flood disasters, but they are by no means comprehensive, or all focused on levee-related specific flood disasters. At some point in the future, when a comprehensive inventory of levees has been completed nationwide and other data becomes available, specifically including costs of levee failures, it will be possible to perform detailed loss of life and economic analyses that would further thoroughly justify the budgets of the NLSP.

The Committee recommends that the Commission:

- Coordinate with all federal, state, and local agencies and other organizations to make sure all existing data has been analyzed;
- Coordinate, transfer, and manage important levee-related flood disaster data within the National Levee Database (NLD); and
- Require that the state and national levee safety programs develop improved methods for tracking damages and avoided costs, and to find improved ways of documenting and disseminating this information.

The Committee also recommends measures to require all federal, state, and local agencies and other organizations coordinate with and provide any available levee-related flood disaster data available to the Commission.

Aerial view of flooding. Pocahontas, AR— Photo by Elmo Webb, PE 3/23/08.

Improved Information Leads to Better Investments

As we look at the historical cost we must also evaluate how risks evolve and compound over time and in turn, impact future costs. The evaluation of risks for the future has various dimensions: (1) the changing landscape due to climate change and subsidence; (2) the changing likelihood of natural hazards such as floods; (3) the degradation of infrastructure due to normal environmental factors; and (4) other evolving factors such as state and regional population, local land use, economic activity, and ecosystem affected by levee failures. A separate, yet constant factor contributing to risk is the fact that risk accumulates with time. Even if the annual chance of occurrence is low, sooner or later, it will happen. At the same time, the probability of adverse consequences also increases as the economy and the population continues to grow.

This view is reiterated in the "Status and Trends" document (URS 2007) prepared for California Delta Vision. This document identifies the following "drivers of future change" for the Sacramento-San Joaquin Delta: Subsidence; Global Climate Change—Sea-Level Rise; Regional Climate Change—More Winter Floods; Seismic Activity; Introduced Species; and Population Growth and Urbanization. These broadly stated drivers of change can be expanded and characterized in various ways but many can generally be applied to most others areas of the country. A full range of reliable information is generally not available or adequate to conduct a detailed, quantitative analysis of each of these drivers of future change. However, based on current prevailing thinking there is every reason to believe that disaster assistance and recovery cost will only continue to increase unless the country significantly changes its floodplain management practices at all levels of government.

Investment in a National Levee Safety Program

Key assumptions and approaches used to develop a cost for a NLSP include the following:

- The governance structure of a NLSP includes the Commissioners, the Commission staff, and the travel and per diem expenses of the four advisory committees.
- Estimates for levee inventory and inspection costs were based upon an assumed scope of an additional estimated 100,000 miles of nonfederal levees (federal levees budgeted for separately).
- Cost-sharing was based on the assumption that setting up the NLSP at the federal level and establishing the Commission would be funded exclusively at the federal level. Similarly, in order to complete the initial inventory and inspection of non-federal levees as soon as possible, it is recommended that this activity also be funded exclusively at the federal level. All other activities, including establishing and maintaining state levee safety programs and the National Levee Rehabilitation, Improvement, and Flood Risk Mitigation Fund would be cost-shared.
- Authorities, appropriations, and staffing for existing federal agencies are leveraged to the maximum extent possible and supplemented where required.
- Estimates of costs for a state levee safety program are derived from a comparison of some similar costs and activities within California.
- Professional judgment was used in the many instances where data did not exist.
- The Committee believes that investments from the NLSP to include the Levee Rehabilitation, Improvement, and Flood Risk Mitigation Fund will return several dollars in benefits for every dollar spent. This is supported by the Corps estimates that for every dollar invested in

flood damage reduction projects there is a $6.48 return on that investment in flood damages prevented.

The Committee further recognizes that there may be instances where the return is marginal when only looking at property damage and economic loss, but when taking into consideration risk to loss of life, the investment can still be well justified.

Major Recommended Elements of a National Levee Safety Program	Annual Costs by Implementation Phase	
	Phase I and II (Years 1-5)	Phase III (Steady State)
National Levee Safety Commission	$40 M (100% Federal) (0% Non-Fed)	$41 M (100% Federal) (0% Non-Fed)
State Levee Safety Programs	$113 M (75% Federal) $37 M (25% Non-Fed)	$85 M (50% Federal) $85 M (50% Non-Fed)
SUBTOTAL	$153 M (Federal) $37 M (Non-Fed)	$126 M (Federal) $85 M (Non-Fed)
	$190 M	$211 M
Complete Initial Non-Federal Levee Inventory and Inspection -Inventory -Initial Inspection -Continuing Management of National Levee Inventory and Database	$25 M (100% Federal) $100 M (100% Federal) N/A	N/A N/A $3 M (100% Federal)
TOTAL	$278 M (Federal) $37 M (Non-Fed)	$129 M (Federal) $85 M (Non-Fed)
	$315 M	$214 M
Levee Rehabilitation, Improvement, and Flood Risk Mitigation Fund	$600 M (65% Federal) $323 M (35% Non-Fed)	$1000 M (65% Federal) $538 M (35% Non-Fed)
	$923 M	$1538 M

Note: Non-federal entities sharing costs include States, Regional Agencies, Local Communities, and Levee Owners and Operators.

Figure 18. Estimated Annual Costs of a National Levee Safety Program.

Putting the National Levee Safety Program in Context

The committee found no existing federal programs for which a direct line item comparison was appropriate due to differences in scope and maturity of existing programs. However, a cursory review of fiscal year 2008 budgets published by the Office of Management and Budget (http://www. whitehouse. gov/omb/ budget/fy2008) indicates that the national program administration elements of the recommendation for a NLSP were similar to or lower than budget line items in agencies such as the Nuclear Regulatory Commission, the National Transportation Safety Board and the Consumer Products Safety Commission. The overall annual estimated costs of the NLSP were roughly comparable to the combined program totals for such federal activities as the Nuclear Regulatory Commission and EPA's Clean Water and Drinking Water programs. The majority of the estimated costs for a NLSP pertain to the rehabilitation of deficient levees (to include non-structural measures) and these estimates represent but a small fraction of the nation's infrastructure needs.

Phased Strategic Plan for Implementation

It has taken more than a century of neglect and indifference for our current levee safety challenges to develop and the solutions that are needed cannot simply be put into place overnight. Due to the massive amount of effort in data collection, assessment, education, policies, procedures and management that is now required, it is essential to roll out the NLSP in well-planned phases.

Each phase is intended to build from the data and experience collected in previous phases. In broad terms, the phases recommended below are designed to help the nation act on critical immediate recommendations, begin steps to implement near term recommendations for a NLSP primarily through incentives, while building the foundational strategies for a sustainable program into the future through both incentives and disincentives. These phased actions are expected to overlap.

Top Two Photos- Levee in major urban area. Dallas, TX— Courtesy of City of Dallas Flood Control District Bottom Photo: Golf course levee. Palm Canyon, CA—Courtesy of Riverside County Flood Control and Water Conservation District.

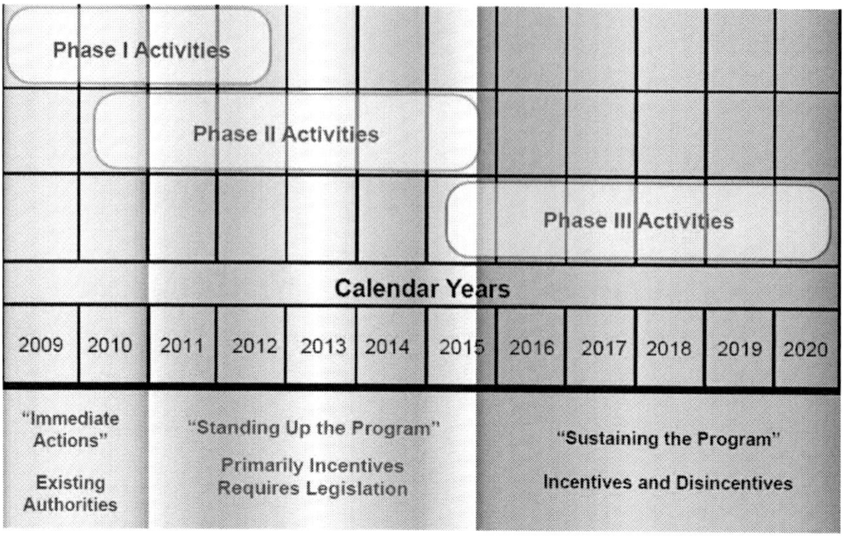

Figure 19. Strategic Implementation of Recommendations on a National Levee Safety Program.

Phase I: Immediate Actions—actions that are time critical and can begin prior to the development of the Commission. Current authorities exist, but funding is needed. Major components include:

1. Congress should pass legislation creating the National Levee Safety Commission (or give authority to existing federal agency).
   a) Appoint Commissioners/Staff Standing Committees
   b) Develop operational plan including legal, technical, financial administrative and institutional procedures

2. Congress should grant authority and appropriations to the Corps to expand the National Levee Database (NLD) and conduct a one-time inventory and inspection program for all levees (federal and non-federal) in the United States.
   a) The Corps should adopt the Interim Hazard Potential Classification System and definitions

3. Congress should fund FEMA to organize a Coordinating Council on Communications for Levees to conduct a needs assessment and begin to develop a public involvement and education/ awareness plan for levee safety.

a) Congress should fund FEMA to develop a Levee Safety website to communicate need for the program, initial risk communication messages and interim technical documents and standards

4. Congress should begin research and implement options to address liability barriers.
5. The International Code Council (ICC) should be employed to develop Interim National Levee Engineering Guidelines.
6. Congress should fund the Corps to begin the Levee Research and Development Program.
7. Congress should mandate risk-based flood insurance behind levees and augment FEMA's mapping program to better communicate risk in living and working behind levees.
8. Change term "levee certification" to "compliance determination."
9. Subject FEMA levee certifications (compliance determinations) to peer review.
10. FEMA and the NFIP Taskforce should explore and implement revisions to CRS Activity 620 to incentivize good levee behavior.

Timing: this phase should begin immediately and run until the Commission is created and fully operational (approximately 2-3 years).

Phase II: Standing Up the National Levee Safety Program—activities designed to create the National Levee Safety Commission, a delegated state program, start-up grant funding and initial incentives. Major components include:

1. Commission should finalize Public Involvement and Education/ Awareness Strategy and Implementation Plan.
2. Operationalize the National Levee Safety Commission (e.g. organization, personnel, guidance, etc.):

a) Develop policies, procedures and guidance for delegated state program;
b) Develop technical materials, direct assistance and training programs including Certified Levee Professional curricula and certification requirements;
c) Administer *National Levee Safety Grant Program* to states;
d) Negotiate with and grant delegation to qualified states; and
e) Begin federal oversight of delegated program.

Legend: = Congress = National Levee Safety Commission

= Federal Government Agencies = States.

Figure 20. Implementation Steps by Actor.

3. Commission should develop and oversee adoption of the National Levee Safety Code through the ICC.
4. Commission should work closely with FEMA and the NFIP Community Rating System Task Force to further explore alignment of FEMA's mitigation grants programs to reward and incentivize good behavior behind levees.

5. Congress/Commission should authorize and fund the National Levee Rehabilitation, Improvement and Flood Mitigation Fund.
6. 6.Commission should develop and implement measures to harmonize levee safety activities with environmental protection requirements.

Timing: this phase should begin as soon as Congress passes legislation to create the National Levee Safety Commission (5-7 years).

Phase III: Sustaining the National Levee Safety Program—activities that result in a mature program, with all needed tools and materials developed. Once this phase is reached, the mix of incentives and disincentives should weigh more heavily towards rewarding superior performers and penalizing states that have not taken action.

1. Commission should finalize the National Tolerable Risk Guidelines for Levees and Canals.
2. Commission should begin to phase in disincentives (e.g. withholding funding for federal programs with a nexus to levee safety) for states that have not developed a state levee safety program.

Timing: this phase should be in place after about 5-10 years.

CLOSING

We are at a critical juncture in our nation's history—a burgeoning growth of risk to people and infrastructure as a result of more than 100 years of inattention to levee infrastructure combined with an economy and social fabric that are vulnerable to catastrophes. The current levee safety reality for the United States is stark—an uncertainty in location, performance and condition of levees and a lack of oversight, technical standards, and effective communication of risks. A National Levee Safety Program is a reasonable and prudent investment that turns the tide on risk growth.

We recognize the need for actions outside of the scope of this report: a broader national flood risk management approach; the benefits of integrating national dam safety and levee safety programs; and leveraging levee safety as a critical first step in a national infrastructure investment strategy. The specific recommendations for a National Levee Safety Program embrace three main concepts:

1) The need for leadership via a National Levee Safety Commission that provides for state delegated programs, national technical standards, risk communication, and collaboration on environmental and safety concerns;
2) The building of strong levee safety programs in all states that in turn provide oversight, regulation, and critical levee safety processes; and
3) A foundation of well-aligned federal agency programs and processes including an initial inventory and inspection of all levees, resolution of liability concerns, and robust incentives and disincentives to stand-up state programs and remediate levee risks.

The Committee recommends a phased strategic implementation with a critical first step to immediately implement Congressional and federal agency actions including legislation establishing a National Levee Safety Program, completion of an inventory and initial inspection of all levees, establishing a Coordinating Council on Communication for Levees, requiring mandatory risk-based flood insurance purchase behind levees, and addressing barriers associated with levee liability. Other phases of implementation will necessarily take years of focused effort to counter the century of inattention.

Now is the time to move thecountry away from a reactive disaster assistance environment to a proactive, safety oriented culture where the general public and governments are informed and able to participate in shared responsibilities of risk management and where levees are reliable. In the post-Katrina environment, we have a clear call to action justified by both improved public safety and smart investment returns. Levee safety deserves a priority focus within national infrastructure needs as levees protect much of the other infrastructure— such as roads, bridges, schools, and water and sewer treatment plants— from frequent flooding.

Floodwall at Industrial Canal. Lower Ninth Ward. New Orleans, LA. 2008—Courtesy of FEMA.

We view the report as a beginning, not an end, to addressing the issue of levee safety and eagerly anticipate the continued dialogue and action regarding the recommendations in the report. Our vision—an involved public and reliable levee systems—finds its refuge in a National Levee Safety Program.

APPENDIX A— NATIONAL COMMITTEE ON LEVEE SAFETY MEMBERSHIP AND CHARTER

DEPARTMENT OF THE ARMY
OFFICE OF THE ASSISTANT SECRETARY
CIVIL WORKS
108 ARMY PENTAGON
WASHINGTON DC 20310-0108

SEP 16 2008

MEMORANDUM FOR DIRECTOR OF CIVIL WORKS

SUBJECT: Implementation of Section 9003, Committee on Levee Safety, of the Water Resources Development Act of 2007

1. The purpose of this memorandum is to provide implementation guidance for Section 9003, titled Committee on Levee Safety, of the Water Resources Development Act (WRDA) of 2007.

2. Section 9003 provides authority to establish a sixteen member "Committee on Levee Safety," with the Chairperson named as the Secretary of the Army. The Committee on Levee Safety (Committee) is to develop recommendations for a national levee safety program, including a strategic implementation plan. Recommendations shall address the nine program goals named in Section 9003. The final report shall be submitted to the Committee on Transportation and Infrastructure of the House of Representatives and the Committee on Environment and Public Works of the Senate not later than 15 January 2009.

3. I have delegated the Chairmanship of the Committee to the U.S. Army Corps of Engineers (USACE) Director of Civil Works, currently Mr. Steven L. Stockton. Remaining membership include the Administrator of the Federal Emergency Management Agency (FEMA) or the Administrator's designee; eight State representatives, one from each USACE division's area of responsibility; two private sector representatives; two local or regional representatives; and two Indian tribe representatives. Recommended Committee members shall be solicited and selected based on criteria established by the Committee Chairperson. I will review these recommendations and appoint the final Committee members.

4. Concomitant with this implementation guidance, I approve the attached charter thereby establishing the Committee on Levee Safety.

John Paul Woodley, Jr.

Encl

John Paul Woodley, Jr.
Assistant Secretary of the Army
(Civil Works)

Printed on Recycled Paper

COMMITTEE ON LEVEE SAFETY

CHARTER

Purpose:

To develop recommendations for a national levee safety program, including a strategic plan for implementation of the program, within 180 days from the date of the initial appropriations for the Committee on Levee Safety (Committee) meeting. Since the technical correction to Title IX, the National Levee Safety Act of 2007 (Act), dated 15 July 2008 permits use of existing appropriations where available, the submission date to Congress is 15 January 2008. Recommendations shall address the nine program goals named in Section 9003 of the Water Resources Development Act (WRDA) of 2007.

Convening Authority:

The Committee is convened under the authority of Section 9003 of WRDA 2007.

Section 9003 Goals:

1. Ensuring the protection of human life and property by levees through the development of technologically, economically, socially, and environmentally feasible programs and procedures for hazard reduction and mitigation relating to levees.

2. Encouraging use of the best available engineering policies and procedures for levee site investigation, design, construction, operation and maintenance, and emergency preparedness.

3. Encouraging the establishment and implementation of an effective national levee safety program that may be delegated to qualified States for implementation, including identification of incentives and disincentives for State levee safety programs.

4. Ensuring that levees are operated and maintained in accordance with appropriate and protective standards by conducting an inventory and inspection of levees.

5. Developing and supporting public education and awareness projects to increase public acceptance and support of State and national levee safety programs.

6. Building public awareness of the residual risks associated with living in levee protected areas.

7. Developing technical assistance materials for State and national levee safety programs.

8. Developing methods to provide technical assistance relating to levee safety to non-Federal entities.

9. Developing technical assistance materials, seminars, and guidelines relating to the physical integrity of levees in the United States.

Definitions:

1. Levee: The term "levee" is defined as "an embankment, including floodwalls" in which,

   - the primary purpose is to provide hurricane, storm, or flood damage reduction relating to seasonal high water, storm surges, precipitation, and other weather events;
   - normally is subject to water loading for only a few days or weeks during a year; and,
   - does not constitute a barrier across a watercourse, such as a dam.

2. Regulatory Authority over Levee Safety: The regulatory authority refers to the ability to promulgate and enforce regulations for the,

   - design and construction of levees or;
   - inspection of levees or;
   - operation and maintenance of levees or;
   - emergency response associated with levees or;
   - management/analysis of the risk and consequences associated with levees or;
   - repair and rehabilitation of levees or;
   - planning and policy development for flood damage reduction projects.

3. Expertise in Levee Safety: Demonstrates experience in the,

   - design and construction of levees or;
   - inspection of levees or;
   - operation and maintenance of levees or;
   - emergency response associated with levees or;
   - management/analysis of the risk and consequences associated with levees or;
   - repair and rehabilitation of levees or;
   - planning and policy development for flood damage reduction projects.

4. State Representative:

- Employee of a State agency with regulatory authority over the safety of any non-Federal levee in the State.
- Has experience with and responsibility for levee safety public policy development.
- Has expertise in levee safety as described in this Charter.

5. Private Sector Representative: Defined as a person who is not an employee of a Federal, State, local, regional government or Indian tribe, with experience in levee safety.

6. Local or Regional Government Representative: Local or regional government is defined as any local or regional entity that can collect taxes or assessments. This could be a city, county, reclamation district, water district, levee district, etc. that has responsibility for levees.

   - Employee of a local or regional agency, which can collect taxes or assessments, such as, a city, county, reclamation district, water district, or levee district.
   - Has expertise in levee safety as described in this Charter.

7. Indian Tribe Representative:

   - Member or employee of an Indian tribe.
   - Has expertise in levee safety as described in this Charter.

Committee Implementation Groups:

Implementation of Committee work will involve the following groups,

1. Committee Voting Membership is to be to be comprised of the 16 Committee members specified in Section 9003 and appointed by ASA(CW):

   - Chairperson: Secretary of the Army or the Secretary's designee (pursuant to 10 USC 3016(b)(3), the Assistant Secretary of the Army for Civil Works (ASA(CW)) shall act for the Secretary of the Army for the purposes of Section 9003)
   - FEMA Representative: Administrator of FEMA or the Administrator's designee
   - Eight State Representatives (one from each USACE Division's Area of Responsibility)
   - Two Private Sector Representatives
   - Two Local/Regional Representatives
   - Two Indian Tribe Representatives

2. Committee Nonvoting Membership to be comprised of subject matter experts selected by the Chairperson.

3. USACE Support Team to be provided by USACE and will be comprised of a project manager, a facilitator, administrative assistants, and other staff deemed necessary by the Chairperson.

4. Review Team to be comprised of members selected from nominees not selected to be a voting or nonvoting member and other organizations. Final review team members shall be selected by the Chairperson.

Roles and Responsibilities:

1. Chairperson: Presides over the Committee and ensures purpose and goals of the Committee are accomplished. Has the ability to appoint a vice chair of his/her choosing to assume the duties of Chairperson in his/her absence.

2. Voting Member: Attend and participate in all Committee meetings. Is responsible for representing the interests and concerns of the organizations or institutions they represent. If a voting member cannot attend a Committee meeting, that member may send an alternate member in their place; however, the alternate member cannot vote. Voting members and alternates are free to abstain from a determination of consensus for whatever reasons and shall adhere to the Committee's charter and operating procedures.

3. Nonvoting Member: Attend and participate in all Committee meetings as subject matter experts. Provide input into Committee and/or work group products. May not send an alternate member in their place during Committee meetings. Nonvoting members shall adhere to the Committee's charter and operating procedures.

4. Review Team Member: Review and provide comments on Committee products when requested and within the timeframe established by the Chairperson.

5. Project Manager: Member of USACE Support Team to serve as lead project manager for the Committee. Responsible for coordinating all activities related to accomplishing the final strategic implementation plan, such as serving as USACE point-of-contact for Committee members, coordinating with others (internal and external to USACE) as needed to support Committee work, managing the facilitation contract, creating communication process to include central location of strategic plan documents, coordinating the review team, attending all Committee meetings, managing project funding and participating in the formulation of the final strategic plan.

6. Facilitator: Member of USACE Support Team to provide meeting planning, facilitation, and note taking services to ensure productive and useful meetings, which successfully engage Committee members and other attendees to accomplish meeting objectives. In addition, provide technical writing services to capture work completed by the Committee in the format of a quality document

presenting the final recommended strategic plan for a national levee safety program.

7. Administrative Support: Member of USACE Support Team to provide administrative support associated with the Committee, which may include processing travel reimbursement, coordinating logistics, and other duties.

Operating Procedures and Guidelines

1. Procedures: The Committee will develop a set of operating procedures and guidelines to set forth in detail how it shall conduct meetings and accomplish the requirements of this charter. These procedures shall also include a communication plan, both internal and external to the Committee.

2. Initial Meeting: The first Committee meeting will be convened in October 2008.

3. Work Groups and Subcommittees: The Committee may create special work groups or subcommittees as necessary to accomplish its purpose. These may include voting and nonvoting members.

4. Meeting Guests: Additional subject matter experts may be invited to attend certain Committee meetings. All guests shall be approved prior to the meeting by the Chairperson.

5. Decision-making: To all extent possible, the Committee's goal is to reach consensus on all substantive issues. Final recommendations of the Committee may be arrived at through consensus among Committee voting members present at a meeting. In cases in which consensus cannot be reached, the Chairperson retains the right to render the recommendations of the Committee at any time. The Chairperson may, at his/her discretion, choose to take a vote from the voting members to inform his/her decision.

6. Charter Amendment: The Committee may propose amendments to the Charter for approval by the ASA(CW).

7. Funding: Voting and nonvoting members will be reimbursed for travel and per diem expenses at rates authorized for an employee of a Federal agency under subchapter I of chapter 57 of title 5, United States Code to accomplish Committee work. USACE will provide resources for the USACE Support Team. The Chairperson can at any time decide to reimburse travel expenses of other participants based on the availability of funds.

8. Term of Appointment: Voting and nonvoting members shall serve an appointment not to exceed two years beginning 1 October 2008. If a voting member notifies the Chairperson he or she is no longer able to serve, the Chairperson may make a recommendation for a replacement in-kind to the ASA(CW) for approval. If a nonvoting member notifies the Chairperson he or she is no longer able to serve, the Chairperson may replace the nonvoting member.

APPENDIX B— NATIONAL LEVEE SAFETY ACT OF 2007

121 STAT. 1288 PUBLIC LAW 110–114—NOV. 8, 2007 National Levee Safety Act of 2007.

TITLE IX—NATIONAL LEVEE SAFETY PROGRAM

33 USC 3301 note. SEC. 9001. SHORT TITLE.

This title may be cited as the “National Levee Safety Act of 2007”.

33 USC 3301. SEC. 9002. DEFINITIONS.

In this title, the following definitions apply:

(1). COMMITTEE.—The term “committee” means the Committee on Levee Safety established by section 9003(a).

(2). INSPECTION.—The term “inspection” means an actual inspection of a levee—

a) to establish the global information system location of the levee;
b) to determine the general condition of the levee; and
c) to estimate the number of structures and population at risk and protected by the levee that would be adversely impacted if the levee fails or water levels exceed the height of the levee.

(3). LEVEE.—

(A) IN GENERAL.—The term “levee” means an embankment, including floodwalls—

I. the primary purpose of which is to provide hurricane, storm, and flood protection relating to seasonal high water, storm surges, precipitation, and other weather events; and
II. that normally is subject to water loading for only a few days or weeks during a year.

(B) INCLUSION.—The term includes structures along canals that constrain water flows and are subject to more frequent water loadings but that do not constitute a barrier across a watercourse.

(4). STATE.—The term “State” means—

a) a State;
b) the District of Columbia;
c) the Commonwealth of Puerto Rico; and
d) any other territory or possession of the United States.

(5). STATE LEVEE SAFETY AGENCY.—The term "State levee safety agency" means the agency of a State that has regulatory authority over the safety of any non-Federal levee in the State.

(6). UNITED STATES.—The term "United States", when used in a geographical sense, means all of the States.

33 USC 3302. SEC. 9003. COMMITTEE ON LEVEE SAFETY.

a) ESTABLISHMENT.—There is established a committee to be known as the "Committee on Levee Safety".

b) MEMBERSHIP.—The committee shall be composed of 16 members as follows:

(1). The Secretary (or the Secretary's designee), who shall serve as the chairperson of the Committee.

(2). The Administrator of the Federal Emergency Management Agency (or the Administrator's designee).

(3). The following 14 members appointed by the Secretary:

a) Eight representatives of State levee safety agencies, one from each of the eight civil works divisions of the Corps of Engineers.

b) Two representatives of the private sector who have expertise in levee safety.

c) Two representatives of local and regional governmental agencies who have expertise in levee safety.

d) Two representatives of Indian tribes who have expertise in levee safety.

(c) DUTIES.—

1) DEVELOPMENT OF RECOMMENDATIONS FOR NATIONAL LEVEE SAFETY PROGRAM.—The committee shall develop recomendations for a national levee safety program, including a strategic plan for implementation of the program.

2) REPORT.—Not later than 180 days after the date of enactment of this Act, the committee shall submit to the Secretary, the Committee on Transportation and Infrastructure of the House of Representatives, and the Committee on Environment and Public Works of the Senate a report containing the recommendations developed under paragraph (1).

(d) PURPOSES.—In developing recommendations under subsection (c)(1), the committee shall ensure that the national levee safety program meets the following goals:

(1). Ensuring the protection of human life and property by levees through the development of technologically, economically, socially, and environmentally feasible programs and procedures for hazard reduction and mitigation relating to levees.

(2). Encouraging use of the best available engineering policies and procedures for levee site investigation, design, construction, operation and maintenance, and emergency preparedness.

(3). Encouraging the establishment and implementation of an effective national levee safety program that may be delegated to qualified States for implementation, including identification of incentives and disincentives for State levee safety programs.

(4). Ensuring that levees are operated and maintained in accordance with appropriate and protective standards by conducting an inventory and inspection of levees.

(5). Developing and supporting public education and awareness projects to increase public acceptance and support of State and national levee safety programs.

(6). Building public awareness of the residual risks associated with living in levee protected areas.

(7). Developing technical assistance materials for State and national levee safety programs.

(8). Developing methods to provide technical assistance relating to levee safety to non-Federal entities.

(9). Developing technical assistance materials, seminars, and guidelines relating to the physical integrity of levees in the United States.

(e)COMPENSATION OF MEMBERS.—A member of the committee shall serve without compensation.

(f)TRAVEL EXPENSES.—To the extent amounts are made available in advance in appropriations Acts, the Secretary shall reimburse a member of the committee for travel expenses, including per diem in lieu of subsistence, at rates authorized for an employee of a Federal agency under subchapter I of chapter 57 of title 5, United States Code, while away from the home or regular place of business of the member in performance of services for the committee.

(g)APPLICABILITY OF FEDERAL ADVISORY COMMITTEE ACT.— The Federal Advisory Committee Act (5 U.S.C. App.) shall not apply to the committee.

33 USC 3303. SEC. 9004. INVENTORY AND INSPECTION OF LEVEES.

(a) LEVEE DATABASE.—

(1). IN GENERAL.—Not later than one year after the date of enactment of this Act, the Secretary shall establish and maintain a database with an inventory of the Nation's levees.

(2). CONTENTS.—The database shall include—

a) location information of all Federal levees in the Nation (including global information system information) and, for non-Federal levees, such information on levee location as is provided to the Secretary by State and local governmental agencies;
b) utilizing such information as is available, the general condition of each levee; and
c) an estimate of the number of structures and population at risk and protected by each levee that would be adversely impacted if the levee fails or water levels exceed the height of the levee.

(3) AVAILABILITY OF INFORMATION.—

a) AVAILABILITY TO FEDERAL, STATE, AND LOCAL GOVERNMENTAL AGENCIES.—The Secretary shall make all of the information in the database available to appropriate Federal, State, and local governmental agencies.
b) AVAILABILITY TO THE PUBLIC.—The Secretary shall make the information in the database described in paragraph (2)(A), and such other information in the database as the Secretary determines appropriate, available to the public.

(b) INVENTORY AND INSPECTION OF LEVEES.—

(1) FEDERAL LEVEES.—The Secretary, at Federal expense, shall establish an inventory and conduct an inspection of all federally owned and operated levees.

(2) FEDERALLY CONSTRUCTED, NONFEDERALLY OPERATED AND MAINTAINED LEVEES.—The Secretary shall establish an inventory and conduct an inspection of all federally constructed, non-federally operated and maintained levees, at the original cost share for the project.

(3) PARTICIPATING LEVEES.—For non-Federal levees the owners of which are participating in the emergency response to natural disasters program established under section 5 of the Act entitled "An Act authorizing the construction of certain public works on rivers and harbors for flood control, and for other purposes", approved August 18, 1941 (33 U.S.C. 701n), the Secretary shall establish an inventory and conduct an inspection of each such levee if the owner of the levee requests such inspection. The Federal share of the cost of an inspection under this paragraph shall be 65 percent.

33 USC 3304. SEC. 9005. LIMITATIONS ON STATUTORY CONSTRUCTION.

Nothing in this title shall be construed as— employees for the recovery of damages caused by an action or failure to act; or

(1). creating any liability of the United States or its officers or employees for the recovery of damages caused by an action or failure to act; or

(2). relieving an owner or operator of a levee of a legal duty, obligation, or liability incident to the ownership or operation of a levee.

33 USC 3305. SEC. 9006. AUTHORIZATION OF APPROPRIATIONS.

There is authorized to be appropriated to the Secretary to carry out this title $20,000,000 for each of fiscal years 2008 through 2013.

Nancy Pelosi
Speaker of the House of Representatives.

Robert C. Byrd
President of the Senate pro tempore.

IN THE HOUSE OF REPRESENTATIVES, U.S.

November 6, 2007.

The House of Representatives having proceeded to reconsider the bill (H.R. 1495) entitled "An Act to provide for the conservation and development of water and related resources, to authorize the Secretary of the Army to construct various projects for improvements to rivers and harbors of the United States, and for other purposes", returned by the President of the United States with his objections, to the House of Representatives, in which it originated, it was Resolved, That the said bill pass, two-thirds of the House of Representatives agreeing to pass the same.

Lorraine C. Miller
Clerk

I certify that this Act originated in the House of Representatives.
Lorraine C. Miller
Clerk.

IN THE SENATE OF THE UNITED STATES,

November 8, 2007.

The Senate having proceeded to reconsider the bill (H.R. 1495) entitled "An Act to provide for the conservation and development of water and related resources, to authorize the Secretary of the Army to construct various projects for improvements to rivers and harbors of the United States, and for other purposes", returned by the President of the United States with his objections, to the House of Representatives, in which it originated, and passed by the House of Representatives on reconsideration of the same, it was Resolved, That the said bill pass, two-thirds of the Senators present having voted in the affirmative.

Nancy Erickson
Secretary.

LEGISLATIVE HISTORY—H.R. 1495 (S. 1248):
HOUSE REPORTS: Nos. 110–80 (Comm. on Transportation and Infrastructure) and 110–280 (Comm. of Conference).
SENATE REPORTS: No. 110–58 accompanying S. 1248 (Comm. on Environment and Public Works).
CONGRESSIONAL RECORD, Vol. 153 (2007):
Apr. 19, considered and passed House.
May 14–16, considered and passed Senate, amended. Aug. 1, House agreed to conference report.
Sept. 24, Senate agreed to conference report.
WEEKLY COMPILATION OF PRESIDENTIAL DOCUMENTS, Vol. 43 (2007):
Nov. 2, Presidential veto message.
CONGRESSIONAL RECORD, Vol. 153 (2007): Nov. 6, House overrode veto.
Nov. 8, Senate overrode veto.

APPENDIX C— ABBREVIATIONS AND ACRONYMS

AAA	Army Audit Agency
ALARP	"As Low As Reasonably Practicable"

ANCOLD	Australian National Committee on Large Dams
ASCE	American Society of Civil Engineers
ASDSO	Association of State Dam Safety Officials
ASFPM	Association of State Floodplain Managers
BIA	Bureau of Indian Affairs
CFR	Code of Federal Regulation
CLP	Certified Levee Professional
COG	Councils of Government
Corps	US Army Corps of Engineers
CRS	Community Rating System
DFIRM	Digital Flood Insurance Rate Map
DHS	Department of Homeland Security
EC	Engineer Circular
ECB	Engineering and Construction Bulletin
ER	Engineer Regulation
ERDC	Engineer Research and Development Center (USACE)
FEMA	Federal Emergency Management Agency
FERC	Federal Energy Regulatory Commission
FIRM	Flood Insurance Rate Map
FY	Fiscal Year
HEC	Hydrologic Engineering Center (USACE)
HEM	Helicopter Electromagnetic
HMGP	Hazard Mitigation Grant Program
HR	House Resolution
HUD	US Department of Housing and Urban Development
IA	Individual Assistance
IBWC	International Boundary and Water Commission
ICOLD	International Commission on Large Dams
MT	Mitigation
NAFSMA	National Association of Flood and Stormwater Management Agencies
NCLS	National Committee on Levee Safety
NFIP	National Flood Insurance Program
NLD	National Levee Database
NLSA	National Levee Safety Act
NLSB	National Levee Safety Board
NLSP	National Levee Safety Program
NOAA	National Oceanic and Atmospheric Administration
NRCS	Natural Resource Conservation Service

O&M	Operations and Maintenance
PE	Professional Engineer
PG	Professional Geologist
PL	Public Law
PSA	Public Service Announcement
R&D	Research and Development
RiskMAP	Risk Mapping, Assessment, and Planning (FEMA)
SES	Senior Executive Staff
TRG	Tolerable Risk Guidelines
US	United States
USACE	United States Army Corps of Engineers
USBR	United States Bureau of Reclamation
USEPA	United States Environmental Protection Agency
USFWS	United States Fish and Wildlife Service
USGS	United States Geological Survey
USSD	United States Society on Dams
WRDA	Water Resources Development Act

APPENDIX D— NATIONAL COMMITTEE ON LEVEE SAFETY RECOMMENDATION DEVELOPMENT AND PUBLIC REVIEW PROCESS

The purpose of this appendix is to describe the process the National Committee on Levee Safety (Committee) followed to formulate the recommendations and solicit feedback from a broad group of organizations and stakeholders.

Committee Member Selection

The Chairmanship of the Committee was delegated to the US Army Corps of Engineers (USACE) Director of Civil Works by the Assistant Secretary of the Army for Civil Works (ASA(CW)). The ASA(CW) selected and appointed final Committee members based on recommendations presented by USACE. Selections were based on criteria, which focused on professional expertise, technical background, leadership and communication experience.

Committee members were charged to bring forth their individual expertise and judgment, and not the views of their organizations. The final recommendations comprise a collaborative Committee product that focuses on national solutions and may not represent the positions of individual members or their organizations.

Committee Operating Framework

The Committee worked intensely from October 2008 to mid-January 2009 through a combination of full Committee meetings, smaller working group meetings, review team meetings, and conference calls. See the

Committee charter, Appendix A, for more details on the operating procedures for the Committee. The following was the schedule:

- Initial committee meeting 6 Oct – 10 Oct 08
- 2nd Committee Meeting 20 Oct – 24 Oct 08
- Review Team Meeting 30 Oct 08
- 3rd Committee Meeting 4 Nov – 8 Nov 08
- 4th Committee Meeting 17 Nov – 21 Nov 08
- 5th Committee Meeting 8 Dec – 12 Dec 08
- Review Team Meeting 12 Dec 08
- Public Webinar 16 Dec 08
- 6th Committee Meeting 5 Jan – 9 Jan 09
- Submit Report 15 Jan 09

Committee members were divided into four work focus groups divided by the goals identified in the National Levee Safety Act. To ensure progress, individual workgroups met regularly at the discretion and organization of workgroup leaders. The following are the workgroups:

Workgroup 1: Technical Assistance (Goals 2, 7, 8, 9) Workgroup 2: Public Awareness (Goals 5, 6)

Workgroup 3: Levee Safety Program Development (Goals 1, 4)

Workgroup 4: Implementation (Goal 3 and linking all other goals)

The Committee followed the following basic steps in its deliberations from October 6, 2008 through January 9, 2009. Because of the compressed timeframe, at times, some of these steps were being conducted in parallel.

Step One: Workgroups developed scoping and clarifying questions for each of the nine goals. Committee presented scoping and clarifying questions for Review Team input.

Step Two: Workgroups identified available data, input and advice needed for formulation of recommendations.

Step Three: Committee conducted field trips to flood damaged areas, levees and appurtenant works in New Orleans, solicited presentations from a variety of experts and consulted technical, scientific and policy documents (for a list of major presenters and documents consulted, see Appendix G).

Step Four: Workgroups developed recommendations for discussion at the plenary that included main steps, rationale, timing, funding, governance, authorities and leverage/impacts on other programs.

Step Five: Committee created a table that mapped recommendations by goal to ensure each goal had been addressed adequately.

Step Six: Committee analyzed, discussed, amended and finalized recommendation content and overall implementation steps.

Review and Feedback Process

Within the constraints of the schedule, the Committee gathered information and feedback from a diverse group of experts and stakeholders throughout the development of the recommendations. Specific activities included the continuous posting of products as they were developed on the Committee website (http://www.iwr.usace.army. mil/ncls); forming a review team and conducting two review team meetings; and hosting a web-based open stakeholder meeting. Committee members reviewed and considered all comments submitted.

The review team was composed of numerous representatives from a range of organizations and interests to serve on the review team. Organizations nominated and/or invited to participate are listed on the following page. Most of the review team members participated in the review meetings in person. Web-based technology was provided for those who chose to participate virtually. Review team members provided verbal and written feedback. Approximately 500 comments were received from the October meeting and approximately 600 comments were received from the December meeting.

The Committee also conducted a two-hour virtual stakeholder meeting on December 16, 2008, to share preliminary recommendations and engage a broader group in a dialogue about the recommendations. This meeting was announced through a media roundtable, all US Army Corps of Engineers public affairs offices and existing professional networks. Approximately 320 individuals participated. Within the time allowed, 22 questions were submitted electronically. A feedback form was sent to all stakeholders to solicit additional comments.

Invited Review Organizations

- American Council of Engineering Companies
- American Public Works Association
- American Rivers
- American Society of Civil Engineers
- American Water Resources Association
- Association of State Dam Safety Officials
- Association of State Floodplain Managers
- Central Valley Flood Protection Board
- Commonwealth of Pennsylvania
- Confederated Salish and Kootenai Tribes
- Department of Transportation
- Federal Emergency Management Agency
- Federal Energy Regulatory Commission
- Federal Highway Administration
- Flood Control District of Maricopa County
- GEI Consultants
- HDR, Inc.
- Hidalgo County Drainage District (TX)
- Institute for Business and Home Safety
- International Boundary and Water Commission
- Klinger and Associates, P.C.
- Middle Rio Grande Conservancy District
- Mississippi River Commission
- National Association of Flood and Stormwater Management Agencies
- National Emergency Management Association
- National Ocean Service
- National Park Service
- National Weather Service

- National Wildlife Federation
- Natural Resource Conservation Service
- Office of Management and Budget, Water and Power Branch
- Ohio Department of Natural Resources, Division of Water
- Pennsylvania Department of Conservation and Natural Resources
- Sacramento Area Flood Control Agency
- Seminole Tribe of Florida
- Small Business Administration
- State of Kansas
- State of Louisiana
- Tennessee Valley Authority
- Terracon Consultants, Inc.
- The Nature Conservancy
- US Army Corps of Engineers
- US Bureau of Indian Affairs
- US Bureau of Reclamation
- US Department of Interior
- US Department of Transportation
- US Environmental Protection Agency
- US Fish and Wildlife Service
- US Forest Service
- US Geological Society
- US Housing and Urban Development
- US Small Business Administration
- US Society of Dams

Appendix E— Applicable Related US Army Corps of Engineers and Federal Emergency Management Agency Programs, Authorities, and Activities

1. Significant Events and Federal Legislation

- Swamp Land Acts 1849, 1850
  - Transferred swamp and overflow land to States on condition that sales revenue was used to build levees

- 1874 Mississippi River Flooding
  - Major flooding on Lower Mississippi resulted in congressional funding for Corps of Engineers study. Study concluded that most ongoing flood control efforts were uncoordinated and inadequate
- 1879 Mississippi River Commission Established
  - Focus was navigation improvements
  - Purpose: Identify and implement the most satisfactory flood control plan possible to improve navigation
- 1917 Flood Control Act
  - First Federal Flood Control Legislation
  - Recognized the federal governments limited responsibilities for flood control in lower Mississippi & Sacramento Rivers
  - Established first cost sharing policy ($2 federal to $1 local)
- 1927 Rivers & Harbor Act
  - Authorized the Corps to conduct surveys of most of the navigable streams of the United States
  - Known as 308 reports they became basic river planning documents
- 1928 Flood Control Act
  - Expanded flood control policy on the Mississippi to include floodways, spillways and channel improvements
  - Released lower Mississippi residents from some local cooperation requirements.
- 1936 Flood Control Act
  - Recognized that flood control was a "proper activity of the federal government in cooperation with states and their localities"
  - Stipulated that federal government would not participate in any flood control project if benefits did not exceed costs.
  - Authorized $320 million for over 200 flood control projects
- Flood Control Act of 1941
  - Section 5 provided authorization to conduct rescue work and repair or maintenance of flood control works threatened or destroyed by flood.
- Emergency Flood Control Act of 1955 (PL 84-99)
  - Created the first authorization for emergency flood response.

(1955) Category 100, 200, 300

(1962) Category 300 HSPP

(1974) Category 400 Contaminated Water Supply

(1976) Category 500 Advance Measures (1977) Category 400 Drought Response (1979) Category 600 Hazard Mitigation (1986) Category 200 Post Flood Response

(1990) Expanded Preparation to "All Natural Hazards"

2. Public Law 84-99

The U.S. Army Corps of Engineers has vested authority under Public Law 84-99 (PL 84-99), as amended, to conduct emergency preparation and response activities to assist public agencies in responding to flood and other emergencies. Assistance can be in the form of technical assistance, direct assistance, or rehabilitation of federal and certain non-federal flood control works damaged or destroyed by floods. Types of assistance are disaster preparedness, advance measures, emergency assistance, flood response, post-flood response, and project rehabilitation. USACE assistance must be requested through the State's Standardized Emergency Management System and coordinated through the State's Response Information Management System. The local agency requesting assistance must provide appropriate documentation (e.g., hold harmless agreements, etc.) following any verbal authorization. FEMA may also assign USACE flood emergency response activities under the Federal Response Plan separately from any PL 84-99 authorization.

3. Water Resource Development Act of 1986 (Public Law 99-662) – Flood Control Act

The major significance of WRDA 1986 was establishing a stronger flood risk reduction sponsor partnership with cost sharing and project development:

- Section 104 – Authority for crediting sponsors for certain work compatible with a federal flood risk reduction project
- Section 204 – Authorizes reimbursement to nonfederal sponsors for construction of authorized federal harbor projects
- Section 902 – Established a twenty percent cap on project cost increases

4. Inspection of Completed Works (ICW) and Rehabilitation and Inspection Program (RIP)

ICW is a Corps of Engineers program that includes periodic inspection of projects. These projects fall under Engineering Regulation (ER) 1110-2-530.

RIP is a Corps of Engineers program to perform inspections of non-federal projects under ER500-1 and the provisions of Public Law 84-99, if requested by the local sponsor.

An initial eligibility inspection must be performed by the Corps of Engineers and subsequent maintenance inspections are required.

Through the Inspection of Completed Works (ICW) and the Rehabilitation and Inspection Program (RIP), the Corps of Engineers performs inspections of flood damage reduction projects, including: (a) projects federally built and maintained; (b) projects federally built and locally maintained; and (c) those projects locally built and maintained to determine eligibility for inclusion in the RIP or to determine eligibility to remain in the RIP. In most cases, maintenance of levees is a local responsibility with oversight provided by the Corps Inspection Program. Levee owners have an incentive to maintain levees in a sound condition to remain in the program and receive rehabilitation assistance after flood events. Additionally, the failure to maintain a levee in sound condition may result in withdrawal of Corps certification that it meets the Federal Emergency Management Agency (FEMA) Base-flood requirement. These inspections are visual verifications of the local entity's compliance with the Operation and Maintenance Manuals and do not include the engineering assessments needed to verify project performance or stability. Results of the inspections are forwarded to the local entity with recommendations for correcting any deficiencies identified.

5. Continuing Authorities Program (CAP)

The Continuing Authorities Program (CAP) is a group of legislative authorities that give the Corps of Engineers the authority to plan, design, and construct certain types of water resources and ecosystem restoration projects without additional and specific Congressional authorization. The purpose of CAP is to implement projects of limited scope and complexity. Each authorityhas specific implementation guidelines, total program and per-project funding limits, and cost share requirements. The following are the most commonly used CAP authorities:

- Small Flood Control Projects authorized by Section 205 of the 1948 Flood Control Act, with a per-project federal funding limit of $7 million. This program is designed to implement projects that reduce overland flood damages. Projects must be technically sound, economically justified and environmentally acceptable.

- Emergency Stream Bank Protection Projects authorized by Section 14 of the 1946 Flood Control Act with a per-project federal funding limit of $1.5 million. These projects are designed protect essential public facilities threatened by flood induced erosion.
- Aquatic Ecosystem Restoration authorized by Section 206 of the 1996 Water Resources Development Act (WRDA) with a per-project federal funding limit of $5 million. This program is designed to develop aquatic ecosystem restoration and protection projects that improve the quality of the environment, are in the public interest, and are cost-effective.
- Project Modifications for the Improvement of the Environment authorized by Section 1135 of the 1986 WRDA with a per-project federal funding limit of $5 million. This program is designed to modify existing Corps projects for the purpose of improving environmental quality.

6. Planning Assistance to States (PAS)

Section 22 of the Water Resources Development Act (WRDA) of 1974 (Public Law 93-251), as amended, provides authority for the Corps of Engineers to assist the states, local governments, and other non-federal entities in the preparation of comprehensive plans for the development, utilization, and conservation of water and related land resources. Section 208 of the WRDA of 1992 (Public Law 102-580) amended the WRDA of 1974 to include eligible Native American Indian tribes as equivalent to a state. Section 2013 of the WRDA of 2007 increased the annual program funding limits to $5 million nationally, with up to $2 million per state or tribe.

The needed planning assistance is determined by the individual states and tribes. Study costs are shared equally by the federal government and the sponsor. Every year, each state and eligible Native American tribe provides the Corps of Engineers its request for studies under the program, and the Corps then accommodates as many studies as possible within the annual funding allotment. Typical studies are only at the planning level of detail; they do not include detailed design for project construction. The studies generally involve the analysis of existing data for planning purposes using standard engineering techniques, although some data collection is often necessary.

The program can encompass many types of studies dealing with water resource issues. Types of studies include the following:

- Water Supply and Demand
- Water Conservation
- Water Quality
- Environmental Conservation and Restoration - Wetlands Evaluation
- Dam Safety/Failure
- Flood Risk Reduction
- Floodplain Management
- Coastal Zone Management and Protection - Harbors and Ports

7. Floodplain Management Services (FPMS)

The program's authority stems from section 206 of the 1960 Flood Control Act (Public Law 86-645), as amended.

Its objective is to foster public understanding of the options for dealing with flood hazards and to promote prudent use and management of the nation's floodplains. Land use adjustments based on proper planning and the employment of techniques for reducing flood damages provide a rational way to balance the advantages and disadvantages of human settlement on floodplains. These adjustments are the key to sound floodplain management. People who live in the floodplain need to know about the flood hazard and the actions that they can take to reduce property damage and prevent the loss of life from floods. The FPMS program was developed by the Corps of Engineers specifically to address this need.

The FPMS programs provide the full range of technical services and planning guidance that is needed to support effective floodplain management. The Technical Services program develops or interprets site-specific data on obstructions to flood flows, flood formation and timing, flood depths, floodwater velocities, and the extent, duration, and frequency of flooding. The Special Studies Program provides assistance and guidance on all aspects of floodplain management planning:

- Floodplain Delineation and Flood Hazard Evaluation - Dam Break Analysis
- Hurricane Evacuation
- Flood Warning and Preparedness
- Regulatory Floodway
- Comprehensive Floodplain Management
- Flood Risk Reduction - Urbanization Impacts - Storm Water Management

- Non-structural Flood Proofing
- Inventory of Flood Prone Structures

Program services are provided without charge upon request to state, regional, and local governments, eligible Native American Indian tribes, and other non-federal public agencies. These entities may provide voluntary contributions toward requested services to expand the scope or accelerate the provision of those services. Program services are also offered to non-water resource federal agencies and to the private sector on a 100 percent cost recovery basis. The Corps has very limited circumstances under which it can accept sponsor funds since the passage of the Thomas Amendment in Section 211 of the WRDA of 2000.

8. National Levee Database Authority (Public Law 109- 148)

Emergency supplemental funds appropriated under Public Law 109-148 (enacted on December 30, 2005) included $30 million for the Corps of Engineers to initiate a National Inventory of Flood and Storm Damage Reduction projects, including an assessment of the condition of levee projects. In addition, the President's budget for Fiscal Year 2007 included $20 million to continue this effort. The Corps is working with FEMA to coordinate its efforts with the FEMA Map Modernization program. It is envisioned that data from the inventory will be able to provide technical information to perform or be used as a basis for periodic re-certification of levees as required by FEMA for floodplain mapping purposes. The inventory will be a geospatial database that will allow data to be incorporated into the flood maps prepared by FEMA or, if more detailed mapping is available, could be used with that mapping. The database will allow users to have real time information readily available.

The Corps completed an initial survey of federal program levee systems in July 2006 and developed a national database to capture information about each levee, including the location and last recorded inspection rating. The levees included in this initial survey are: (1) federally owned and maintained; (2) federally built and locally maintained; and (3) locally built and maintained that meet specified Corps standards. The initial Corps survey included approximately 2,000 levees, encompassing approximately 13,000 miles, in the Corps Inspection of Completed Works (ICW) and Rehabilitation and Inspection (RIP) programs. Many of these projects were authorized by Congress for federal construction and later turned over to state and local sponsors to operate and maintain. These projects are inspected on a biennial schedule.

9. Water Resources Development Act of 2007 (Public Law 110-114) – National Levee Safety Act of 2007

This WRDA established the National Committee on Levee Safety (NCLS) and charged it with developing a national levee safety policy. Section 9003 of Title IX listed nine areas of concern to be addressed by the NCLS in a report to Congress:

1) Ensuring the protection of human life and property by levees through the development of technologically, economically, socially, and environmentally feasible programs and procedures for hazard reduction and mitigation relating to levees.
2) Encouraging use of the best available engineering policies and procedures for levee site investigation, design, construction, operation and maintenance, and emergency preparedness.
3) Encouraging the establishment and implementation of an effective national levee safety program that may be delegated to qualified States for implementation, including identification of incentives and disincentives for State levee safety programs.
4) Ensuring that levees are operated and maintained in accordance with appropriate and protective standards by conducting an inventory and inspection of levees.
5) Developing and supporting public education and awareness projects to increase public acceptance and support of State and national levee safety programs.
6) Building public awareness of the residual risks associated with living in levee protected areas.
7) Developing technical assistance materials for State and national levee safety programs.
8) Developing methods to provide technical assistance relating to levee safety to non-Federal entities.
9) Developing technical assistance materials, seminars, and guidelines relating to the physical integrity of levees in the United States.

Section 9004 of Title IX The legislation also expanded the National Levee Database from listing federal levees to include all levees in the United States, with an emphasis on condition, establishing the population at risk and determining location by GIS coordinates.

Section 9006 of Title IX authorized $20 million per year for each federal fiscal year from 2008 through 2013.

The Federal Emergency Management Agency (FEMA) Authorities and Activities:

A. Statutes/Legislation:

3) National Flood Insurance Program (NFIP): Includes flood hazard identification (mapping, including areas impacted by levees), floodplain management, and flood insurance authorities.The U.S. Congress established the National Flood Insurance Program (NFIP) with the passage of the National Flood Insurance Act of 1968. The NFIP is a Federal program enabling property owners in participating communities to purchase insurance as a protection against flood losses in exchange for State and community floodplain management regulations that reduce future flood damages. Participation in the NFIP is based on an agreement between communities and the Federal Government. If a community adopts and enforces a floodplain management ordinance to reduce future flood risk to new construction in floodplains, the Federal Government will make flood insurance available within the community as a financial protection against flood losses. This insurance is designed to provide an insurance alternative to disaster assistance to reduce the escalating costs of repairing damage to buildings and their contents caused by floods. Legislation relating to the NFIP include:

I. The National Flood Insurance Act of 1968 (NFIA)
II. The Flood Disaster Protection Act of 1973 (FDPA)
III. The National Flood Insurance Reform Act of 1994 (NFIRA 1994)

- Resulted in major changes to the NFIP. NFIRA, which amended the FDPA, provides tools to make the NFIP more effective in achieving its goals of reducing the risk of flood damage to properties and reducing Federal expenditures for uninsured properties that are damaged by floods.
- Community Rating System –Subtitle C Section 541. Community Rating System and Incentives for Community Floodplain Management.
  - To provide incentives for measures that reduce the risk of flood or erosion damage that exceed the criteria set forth in Section 1361 and evaluate such measures;
  - To encourage adoption of more effective measures that protect natural and beneficial floodplain functions;
  - To encourage floodplain and erosion management; and

  - o To promote the reduction of Federal flood insurance losses.
- Flood Mitigation Assistance Grant Program:
  - o Pre-disaster grant program that provides funds every year to states and communities for projects that reduce or eliminate the long-term risk of flood damage to buildings, homes, and other structures that are insured under the NFIP.

IV. The National Flood Insurance Reform Act 2004 (NFIRA 2004): The Bunning-BereuterBlumenauer Flood Insurance Reform Act of 2004 (PL 108–264), which amended the National Flood Insurance Act (NFIA) of 1968 (42 U.S.C. 4001, et al.)

  - o Repetitive Flood Claims Grant Program
  - o A pre-disaster nationally competitive grant program that funds mitigation projects for certain repetitive loss properties in communities or states that cannot participate in the FMA program because they do not have funds for the non-federal match or lack the capacity to manage FMA grant activities. (http://www.fema.gov/government/grant/ rfc/index.shtm)
  - o Severe Repetitive Loss (SRL) Grant Program
  - o A pre-disaster grant program that is
  - o reserved for "severe" repetitive loss properties (i.e., residential properties with a high frequency of losses or a high value of claims). The funding is used to reduce or eliminate the long-term risk of flood damage to SRL structures insured under the NFIP. (http://www.fema.gov/government/ grant/srl/index.shtm)

2) Disaster Assistance:

Robert T. Stafford Disaster Relief and Emergency Assistance Act (the Stafford Act), PL 100-707: Signed into law November 23, 1988; amended the Disaster Relief Act of 1974, PL 93-288. This Act constitutes the statutory authority for most Federal disaster response activities especially as they pertain to FEMA and FEMA programs. The Stafford Act provides the statutory framework for a Presidential declaration of an emergency or a declaration of a major disaster. Such declarations open the way for a wide range of federal resources to be made available to assist in dealing with the emergency or major disaster involved. The Stafford Act structure for the declaration process reflects the fact that federal resources under this statute supplement state and local resources for disaster relief and recovery. Except in the case of an emergency involving a subject area that is exclusively or preeminently in the federal purview, the Governor of an affected state, or Acting Governor if the Governor is not available, must request such a declaration by the President.

Financial Assistance:

I. Individual Assistance

The FEMA Individual and Households Program (IHP) provides assistance to victims of presidentially declared disasters. IHP assistance can be available to individuals, families and businesses. Assistance can include temporary housing, financial assistance for repairing a damaged dwelling, and assistance with other disaster-related needs such as transportation or medical and dental expenses incurred as a result of the disaster. IHP assistance is meant to help those affected by disasters with critical expenses that cannot be covered in other ways; it is not intended to restore an individual's damaged property to its condition before the disaster. While some housing assistance funds are available through the Individuals and Households Program, most disaster assistance from the Federal government is in the form of loans administered by the Small Business Administration.

II. Public Assistance- Section 406 of the Stafford Act

- Public Assistance is a post-disaster program established under Section 406 of the Stafford Act that is jointly administered by FEMA and individual states. As part of the reimbursements made to restore damaged public facilities and certain private nonprofit (PNP) facilities, public assistance funds may be made available for cost-effective mitigation measures undertaken as part of the recovery. The amount of Section 406 Mitigation funds made available in any given disaster is not computed by a formula, but is based on a project-by-project evaluation of the feasibility and cost-effectiveness of mitigation measures.

Post-Disaster Grant Program Assistance:

i. Hazard Mitigation Grant Program (HMGP) – Section 404 of the Stafford Act
   - The Hazard Mitigation Grant Program offers post-disaster funding to states, communities, and other eligible grant recipients to invest in long-term measures that will reduce vulnerability to future natural hazards. The states have a strong role in administering HMGP, with FEMA providing oversight.

ii. Pre-Disaster Mitigation (PDM) – Section 203 of the Stafford Act
   - Pre-Disaster Mitigation is a nationally competitive grant program designed to assist states and communities to develop mitigation

plans and implement mitigation projects. PDM funds are appropriated annually. FEMA convenes national panels to evaluate eligible applications that are submitted by states following the state selection process.

a) Hazard Mitigation Planning

I. Disaster Mitigation Act of 2000 (DMA) (PL 106-390): Amends Section 322 of the Robert T. Stafford Disaster Relief and Emergency Assistance Act (Stafford Act), 42 U.S.C. 5165, and provides for States, Tribes, and local governments to undertake a risk-based approach to reducing risks to natural hazards through mitigation planning. The National Flood Insurance Act of 1968, as amended, 42 U.S.C. 4001 et seq, reinforced the need and requirement for mitigation plans, linking flood mitigation assistance programs to State, Tribal and Local Mitigation Plans.

b) B. Regulations:

1) NFIP:

a. Title 44 Code of Federal Regulations (CFR) Part 65 - Identification and Mapping of Special Hazard Areas

I. 44 CFR 65.10 – Mapping of Areas Protect by Levee Systems – Established on August 25, 1986. FEMA regulatory responsibilities with regard to mapping areas protected by levees. FEMA is charged with accrediting levees certified by others, determining the appropriate flood risk designations for areas behind levees, and accurately depicting these flood risks on flood hazard maps

b. Parts 59, 60, 61, and others covering flood insurance and floodplain management activities

2) Disaster Assistance:

a) Title 44 CFR Part 206 – Federal Disaster Assistance

3) Mitigation Grants:

a) Title 44 CFR Part 79 – Flood Mitigation Grants

4) Hazard Mitigation Planning:

a) Title 44 CFR 201 – Mitigation Planning

c. Policies/Guidance:

1) NFIP:

a. FEMA's Guidelines and Specifications for Flood Hazard Mapping Partners Guidelines and Specifications Appendix H:

i. This Appendix describes the FEMA requirements and procedures for evaluating earthen levee systems and mapping the areas affected by those systems.

b. Procedural Memorandums (PMs): PMs supplement and clarify the information in Appendix H of FEMA's Guidelines and Specifications for Flood Hazard Mapping Partners on mapping the base flood in areas with levees.

i. FEMA Procedural Memo 34 - Interim Guidance for Studies Including Levees Aug. 22, 2005.

- This Procedure Memorandum provides FEMA staff, contactors, and mapping partners with guidance for the evaluation and mapping of levees and levee-affected areas as part of the FEMA Flood Map Modernization effort.

ii. ii.FEMA Procedural Memo 43 - Guidelines for Identifying Provisionally Accredited Levees (revised) Mar. 16, 2007 - Supersedes version issued on Sept. 25, 2006

- This Procedure Memorandum provides FEMA staff, contractors, and mapping partners with guidance for identifying Provisionally Accredited Levees (PALs) and mapping levee-affected areas. Also included is a fact sheet, prepared in question-and-answer format, that provides detailed information regarding National Flood Insurance Program procedures for the evaluation and mapping of levee systems with emphasis on Procedure Memorandum No. 43 and PAL systems. This fact sheet was designed for a more technical audience. Additional documents include flowcharts and sample letters for different levee scenarios.

c. CRS Guidance

2) Disaster Assistance

a. Individual Assistance Policy and Guidance
b. Public Assistance Policy and Guidance
c. Hazard Mitigation Assistance Program Guidance
d. Hazard Mitigation Planning Guidance
c. CRS Guidance

APPENDIX F— CURSORY COST ESTIMATES FOR A NATIONAL LEVEE SAFETY PROGRAM

Cursory Cost Estimates for National Levee Safety Commission Activities

Preliminary costs were estimated for the following components:

- Establishing and maintaining Commission members, staff and Advisory Committees.
- Technical Programs, including establishing National Levee Safety Code, publications, developing and distributing training materials, providing technical assistance, and establishing and maintaining a research and development program.
- Remapping FEMA NFIP maps to establish AL and XL zones, and other augmentations of FEMA mapping programs.
- Leading public involvement and education/awareness campaigns to improve the understandings of risk and to change behavior in leveed areas.
- Developing and implementing measures and practices to more closely harmonize levee safety activities with environmental protection requirements and principles.

Costs were estimated for both a 5-year initial start-up phase, and a steady-state or long-term phase. Average costs for both phases are displayed in Table F-1 below.

Cursory Cost Estimates for State Levee Safety Programs

Recent experience from California was used to estimate the costs necessary for establishing and maintaining an average State Levee Safety Program. The process for this was as follows:

- The first step was to take the estimated 100,000 miles of non-federal levees in the nation and assume that the average state program would involve approximately 2,000 miles of levees. Using the experience from California for 1,600 miles of state-federal project levees, as detailed in Table F-2, it was estimated that there would be an average one-time start-up cost of approximately $6.5 million.After start-up, there would be an average annual cost of approximately $3.4 million.

- Taking the average annual cost of $3.4 million per year per state would end up totaling approximately $170 million per year for 50 states.
- It was assumed that the average one-time startup cost of $6.5 million would be spent over five years. This would lead to a total start-up cost of approximately $65 million per year for 50 states spread over each of the first five years. However, during this same time, some states will have completed portions of their initial start-up activities and begun accruing some of the long-term annual costs. If we assume during the first five years that, on the average, about half of the long-term annual costs are being expended, then the average annual costs for all 50 states during the first five-year startup period would be approximately $150 million [$65 million + (0.5 x $170 million)].
- It was assumed that the average annual cost for all 50 states during the first five years would be cost-shared, with the federal government paying approximately $113 million (75%) and non-federal entities paying approximately $37 million (25%) per year.
- It was assumed that the average annual cost for all 50 states during the long-term steady state phase would also be cost-shared, with the federal government paying approximately $85 million (50%) and non-federal entities paying $85 million (50%) per year. The rationale for the lower federal cost share for the long-term steady state phase is that the costs of continued annual inspections would be expected to be borne completely by the non-federal entities over long-term.

Average costs for both phases are displayed in Table F-2.

Table F-1. Estimated Costs for Establishing and Maintaining a National Levee Safety Commission

National Levee Safety Commission	Annual Costs by Implementation Phase			Annual Costs by Implementation Phase		
	Phases 1 and II (Years 1 – 5)			Phase III (Steady State)		
Activity	*Cost*	*Cost-Share*		*Cost*	*Cost-Share*	
		Federal	*Non-Federal*		*Federal*	*Non-Federal*
Commissioners, Commission Staff, Advisory Committees, and Managing State Program Delegation	$15M	100%	0%	$20M	100%	0%
Technical Programs—Codes, Publications, Training, Technical Assistance, and Research & Development	$11M	100%	0%	$13M	100%	0%
Remapping for AL and XL Zones and augmenting FEMA mapping program	$10M	100%	0%	$5M	100%	0%
Public Involvement and Education	$3M	100%	0%	$2M	100%	0%
Environment and Public Safety	$1M	100%	0%	$1M	100%	0%
TOTAL	$40M			$41M		

Note: Federal funds to assist state levee safety programs are envisioned to flow to the agency that is actually performing the federally funded work. It is intended that much of the funding would be delivered to the responsible agency to perform functions such as inspections, preparation of reports and emergency action planning (see section entitled Strong Levee Safety Programs in All States) for more detail.

Table F-2. Estimated Costs for State Levee Safety Programs

Activity	One-time Cost ($1,000)	Annual Cost ($1,000)	Comments	Basis for Cost
Adopt National Code	$400	$100	Enact regulations, supplement, and update. This can involve significant staff effort and public review.	Assumed
Safety inspections annually	$400	$1,500	Program setup and training would be signficant. Likely more expensive for locals to perform work and state to have some oversight. Assume typical state has 2,000 miles nonfederal levees currently uninspected.	California 1,600 miles cost of $1.5 M for twice annually.Increased by 25% for 2,000 miles. Decreased by 33% for once annually. Increased by 20% for state oversight of local inspection. Assumed cost for initial setup and training.
Identify hazard potential of levees	$20	$20	Some initial training. Includes identifying possible new jurisdictional levees.	Assumed
Provide updated information to NLD	$100	$50	Program setup and training. Some annual maintenance.	Assumed
Emergency action plans and evacuation plans/2,300 miles (includes 300 miles of federal levees)	$4,300	$50	Add 300 miles of federal levee since evacuation plans currently not required for federal levees— for total of 2,300 miles of levees. Assume 500 miles of high hazard potential levees need robust plans and 900 miles of significant hazard potential levees need 1/2 of the effort of a robust plan.	Use Sacramento County as cost basis for robust plans. Sacramento County 2004 cost of $325,000 for 90 miles of levee. Increased by 25% to include some additional effort and inflation. Assumed annual cost of $50,000 for periodic updates.
Enter public or private property for inspection/ response	$0	$0	Would likely take legislation, possibly controversial.Legislative activities already funded.	Assumed

Activity	One-time Cost ($1,000)	Annual Cost ($1,000)	Comments	Basis for Cost
Provide risk notification and public outreach	$300	$500	Could spend much less or more, depending on how thorough the outreach must be. Initial annual cost may start out low, but would expect to increase to something like $500,000 per year. Develop initial communication plan thru public input and research.	Assumed gathering public input and initial modest level of outreach involving public meetings, newspaper ads, PSAs, internet, email. Over time could approach something closer to California's effort. California has budgeted $1 million annually for 1,600 miles of levees, with mailer to every property owner. Assume $500,000 annually for well-developed state outreach program. Assumed cost of initial communication plan and program setup.
Provide reports on program status and performance	$200	$300	Inspection reporting costs are covered above.	Assumed
Promulgate rules and procedures	$800	$100	Enact regulations, supplement, and update. This can involve significant staff effort and public involvement.	Assumed
Consider nonstructural measures		$100	Initial procedures are covered above. Provide guidance and assistance to communities for nonstructural programs.	Assumed
Act as coordinator		$400	Initial procedures are covered above. Provide coordination within state and with national level program on levee safety program issues.	Assumed
Prepare and approve grant application packages		$90	Initial procedures are covered above. Assume this involves verifying that the updated HMPs cover levees.	Assumed 3% administration cost for national grants of $3 million/state.

Table F-2. (Continued)

Activity	One-time Cost ($1,000)	Annual Cost ($1,000)	Comments	Basis for Cost
Receive, disburse, and administer		$210	Initial procedures are covered above.	Assumed 7% administration cost for national grants of $3 million/state.
TOTAL	$6,520	$3,420		

Note: Estimated cost for a state with 2,000 miles of non-federal levees.

APPENDIX G— SOURCES PRESENTED OR CONSULTED BY NATIONAL COMMITTEE ON LEVEE SAFETY (NCLS)

The National Levee Safety Committee actively sought and benefited immensely from its consultation with experts in a variety of disciplines and fields and from the rich history of studies and reports issued previously on the topic of levee safety and floodplain management. Below is a list of this source material that was consulted by the Committee and informed its discussions and recommendations.

Presentations

- ASFPM Foundation Report Levees 2050. Sam Riley Medlock, Association of State Floodplain Managers (ASFPM), October 2008.
- ASFPM/NAFSMA Joint Wye River Levee Policy Summit Recommendations. Susan Gilson (NAFSMA), October 2008.
- California's FloodSAFE Program. Rod Mayer, Assistant Deputy Director, FloodSAFE, State of California, October 2008.
- Congressional Reasearch Service; Teleconference on Governance Issues. Nicole Carter, Claudia Copeland, Mary Tiemann, Jim McCarthy, Rob Meltz, October 30, 2008.
- Dam Safety Program Structure, USACE: Governance and Program Scope Overview. Eric Halpin, Special Assistant for Dam and Levee Safety, USACE, October 2008.

- Double Edged Sword. Chad Berginnis, Association of State Floodplain Mangers (ASFPM), October 2008.
- FEMA Programs as Incentives or Disincentives to National Levee Safety Program. Bill Blanton and Craig Kennedy, FEMA, November 2008.
- FEMA's Programs that Relate to Levees. Bill Blanton, Chief of Engineering and Management, FEMA, October 2008.
- Flood Risk Communication. Mary Jo Vrem (FEMA), teleconference, November 14, 2008.
- A Focus on Behavior Change: Applying social marketing to reducing risks around levees. Peter Mitchell, Marketing for Change, November 2008.
- How We Got Where We are Today: An Historical Perspective on Levees and Summary of Issues. Dr. Gerry Galloway, University of Maryland, October 2008.
- Hurricane Katrina Response and Recovery. James B. Walters, USACE, November 2008.
- Improving Flood Protection—Understanding How Levees are Different from Dams. Les Harder, Senior Water Policy Advisor, HDR, Inc., October 2008.
- Keeping the Strategic in Your Strategic Plan. Philip Rizzi, Business Program Manager, Human Capital Account, SRA International, October 2008.
- Learning from Katrina: Actions for Change and Implementing the IPET Recommendations. Gary House, Actions for Change Program Manager, USACE, October 2008.
- Levee Policy Summits: Outcomes and Summary. Dusty Williams & Susan Gilson, National Association of Flood and Stormwater Management Agencies (NAFSMA), October 2008.
- Levee Safety Act, Title IX Overview. Eric Halpin, Special Assistant for Dam and Levee Safety, USACE, October 2008.
- Mississippi River Commission: History, Organization, Governance and Authorities. Stephen Gambrell, R.D.
- James, Member, and Charles Camillo, Mississippi River Commission (MRC), November 2008.
- New Orleans Hurricane and Storm Damage Risk Reduction System (HSDRRS). Karen Durham Aguilera, Director, Task Force Hope, USACE, November 2008.

- Overview of the Delaware River Commissions Organization and Structure. Carol Collier, Executive Director, Delaware River Commission, November 2008.
- Status of the National Levee Database. Tim Pangburn, Chief of Remote Sensing/GIS and Water Resources Branch ERDC-CRREL, USACE, October 2008.
- Tolerable Risk. Eric Halpin, Special Assistant for Dam and Levee Safety, USACE, October 2008.
- USACE Levee Safety Program. Tammy Conforti, Levee Safety Program Manager, USACE, October 2008.

Relevant Reports and Documents

- American Institutes for Research. 2006. An Evaluation of the National Flood Insurance Program (NFIP) Final Report. Washington, DC.
- Association of State Floodplain Managers. 2003.
- Community Liability and Property Rights.
- Association of State Floodplain Managers. 2006. Liability for Water Control Structure Failure Due to Flooding.
- Association of State Floodplain Managers Foundation.
- 2008. A Comparative Look at the Public Liability for Flood Hazard Mitigation.
- Association of State Floodplain Managers Foundation.
- 2008. Floodplain Management 2050: A Report of the 2007 Assembly of the Gilbert F. White National Flood Policy Forum. Washington, DC.
- Charterizing Flood Risk for More Informed Public Involvement. Version 1. Charles Yoe. September 2006.
- Congressional Research Reports for the People. 2008.
- Federal Liability for Flood Damage Related to Army Corps of Engineers Projects.
- Federal Tolerable Risk Workshop, March 18-19, 2008.
- Alexandria, Virginia.
- FEMA. 1986. A Unified National Program for Floodplain Management. Washington, DC.
- Galloway, Gerald E., Brigadier General, US Army, Executive Director, Interagency Floodplain Management Review Committee,

Washington, DC. June 30, 1994. Sharing the Challenge: Floodplain Management into the 21st Century, A Blueprint for Change.
- Interagency Floodplain Management Review Committee. 1994. Sharing the Challenge: Floodplain Management into the 21st Century. Washington, DC: US Government Printing Office.
- Iteragency Levee Policy Review Committee, Washington, DC. September 2006. The National Levee Challenge; Levees and the FEMA Flood Map Modernization Initiative.
- Interagency Performance Evaluation Taskforce (IPET). 2007. Interagency Performance Evaluation Taskforce (IPET) Final Report, Volume I-VIII.
- Interstate Council on Water Policy. 2008. Interstate Organizations Map. Washington, DC.
- Interstate Council on Water Policy. 2002. Interstate River Basin Organization Source Water Protection Survey. Washington, DC.
- Interstate Council on Water Policy. 2006. Interstate Water Solutions for the New Millennium. Washington, DC.
- Louisiana Coastal Protection and Restoration Authority. 2007. Integrated Ecosystem Restoration and Hurricane Protection: Louisiana's Comprehensive Master Plan for a Sustainable Coast. Baton Rouge, LA.
- Mehan, G. Tracy III. 2008. Let's Drink to Private Water.
- Mehan, G. Tracy III. 2007. Mehan on Water: God Gave Us the Water, but Who Pays for the Pipes? Water & Wastes Digest.
- The National Levee Challenge: Levees and the FEMA Flood Map Modernization Initiative, Report of the Interagency Levee Policy Review Committee (2006).
- Netherlands Water Partnership. 2008. National Levee Design and Monitoring Standards in the Netherlands Presentation.
- Public Hazards Communication and Education: The State of the Art. 2004. Update of Informer Issue 2: Public Education for Earthquake Hazards. Transforming the Corps into a Risk Managing Organization. David Moser, Todd Bridges, Steve Tone, Yocov Haimes, Brian Leonard Shabman, Charles Yoe. undated
- US Army Corps of Engineers, Institute for Water Resources. 2008. Decision-Making Chronology for the Lake Pontchartrain and Vicinity Hurricane Protection Project.
- US Army Corps of Engineers. 2008. Draft: Louisiana Coastal Protection and Restoration Technical Report. New Orleans, LA.

- US Army Corps of Engineers. 2008. Stemming the Chocolate Tide: Documentary about the project flood on the Mississippi River.
- US Department of Transportation Federal Highway Administration. 2008. Memo on Highway Embankments versus Levees and other Flood Control Structures.
- US Environmental Protection Agency (EPA), Office of Water. 2002. The Clean Water and Drinking Water Infrastructure Gap Analysis.

Reference Documents

- FEMA. 2008. Acronyms and Abbreviations: Frequently Used for Levee Systems.
- FEMA. 2008. Frequently Used Terms for Levee Systems.
- FEMA. 2008. Meeting the Criteria for Accrediting Levee Systems on NFIP Flood Maps.
- FEMA. 2008. The NFIP and Levee Systems: An Overview.
- FEMA. 2008. The NFIP and Levee Systems: Frequently Asked Questions.
- FEMA. 2008. Provisionally Accredited Levees.
- FEMA. 2008. Requirements of 44 CFR Section 65.10:
- Mapping of Areas Protected by Levee Systems.
- Halpin, Eric. USACE. 2008. Flood Risk Management Program: Levee Knowledge Session. Washington, DC.
- H.R. 4050. 110th Congress, 1st Session.
- Memorandum of Agreement between the US Army Corps of Engineers and the Ministry of Transport, Public Works, and Water Management of the Netherlands.
- Office of Management and Budget. 1997. Floodplain Management and Procedures for Evaluation and Review of Levee and Associated Restoration Projects.
- Thompson, Robert. ASCE. Standard of Care Versus Codes and Standards. September/October 2008.
- US Army Corps of Engineers. National Dam Safety Program Act (NDSPA).

ACKNOWLEDGMENTS

We would also like to acknowledge the dedicated efforts of the many contributors, presenters, support staff, contractors, and committee preparations for the National Committee on Levee Safety. The following organizations and personnel made this report possible:

SRA International, Inc.—for professional facilitation and support of the National Committee on Levee Safety, especially Linda Manning (facilitator), Laura Sneeringer, Samantha Levine, Nan Thorn-Clements, Shannon Kearney, Philip Rizzi and Peter Trick.

Pittsburgh District, US Army Corps of Engineers, for administrative and financial management assistance, especially Kathleen Takach and Joyce Shanko.

US Army Corps of Engineers, Levee Safety Policy and Procedures Team, especially Michael Bart, Pete Pierce, Tammy Conforti, Meegan Nagy, Jeff Jensen, Tim Pangburn, and Tracy Hendren.

National Committee on Levee Safety Review Team—please see Appendix D for a list of individual and organization reviewers.

The many presenters that were kind enough to share their expertise, experience and perspectives— please see Appendix G for a list of presenters and reports consulted in this effort.

Most importantly, the Committee members would like to thank their families, friends and co-workers for their support during the intense 3-month report development period.

In: National Levee Safety…
Editors: D. Mancuso, V. Masterson
ISBN: 978-1-61470-511-6

Chapter 2

THE CONTEXT: CURRENT STATE OF LEVEES AND PUBLIC SAFETY*

National Committee on Levee Safety

FREQUENTLY ASKED QUESTIONS

The Context: Current State of Levees and Public Safety

1. What Problem is the National Committee on Levee Safety Trying to Address?

We are at a critical juncture in our nation's history – risks of loss of life, property damage, and damage to our natural environment behind levees are increasing. The consequences of levee failure could be devastating. The situation is the result of more than 100 years of inattention to, and in some cases neglect of, levee infrastructure, combined with a growing population living in leveed areas and an economy and social fabric that are in a particularly vulnerable state. Due to decades of inattention and underinvestment, there is not a full understanding of where the risk exists and how significant the impacts of failure might be. Here is what we know:

* This is an edited, reformatted and augmented version of a A Report to Congress from the National Committee on Levee Safety publication, dated February, 2011.

- Much of our levee infrastructure is decades old and was built without the benefit of modern engineering.
- Levees protect other critical infrastructure, such as hospitals, roads, drinking and wastewater facilities, and power generating facilities.
- Development continues to intensify behind levees, putting more people, property, and critical infrastructure at risk.
- In many cases, the scale of development behind levees is outpacing the level of the levee's flood risk reduction.
- Much of the public remains unaware of their risk due to levee failure and overtopping and unaware of the actions they could take to lessen that risk.
- Levee safety must be a shared responsibility requiring action by all levels of government – federal, state, regional, and local – and the levee owners and operators. No single level of government, agency, or organization can overcome these challenges alone.

In addition to reducing the risk of living and working in leveed areas, the National Committee on Levee Safety's recommendations for a National Levee Safety Program are designed to reduce the overall costs of levee-related flooding and make our national policies more aligned and rational. For example, by doing nothing, we continue to pay exorbitant costs in the form of immediate disaster relief and post-disaster rebuilding. The National Committee on Levee Safety (NCLS) estimated economic damages resulting from flood damages in leveed areas to be between $5 and $10 billion annually, based on data from the U.S. Army Corps of Engineers, State of California, and flood insurance data.

The National Committee on Levee Safety

Congress created the National Committee on Levee Safety to develop recommendations for a national levee safety program, including a strategic plan for implementation of the program. The NCLS adopted the vision of an involved public and reliable levee systems working as part of an integrated approach to protect people and property from floods, and has been working toward this goal since October 2008.

The NCLS recommendations for a National Levee Safety Program are based on three central concepts:

- Leadership via a National Levee Safety Commission that provides for participating state levee programs, national technical standards, risk communication, and coordination of environmental and safety concerns;
- Strong levee safety programs in and within all states that, in turn, provide oversight and critical levee safety processes; and
- A foundation of well-aligned federal agency programs and processes.

For more information on the NCLS and its recommendations for a National Levee Safety Program, please visit: http://www.leveesafety.org.

2. If I Don't Live in New Orleans, Should Levee Safety Concern Me?

Levees are built in all 50 states, and can be found in 22% of counties nationwide. The nation's attention was refocused on the role of levees as critical infrastructure with the Mississippi River floods (1993 and 2008), California floods (1986 and 1997), and, of course, the tragedy of Hurricane Katrina (2005).

With economic damages stemming from Hurricane Katrina estimated to be more than $200 billion and a loss of life of more than 1,800 persons, the role of levees in providing for public safety and flood risk management was thrust back into the national spotlight. That experience reminded us of the following:

- What were thought to be good levees and flood walls can fail, and do so catastrophically.
- The complexities associated with communicating risk and evacuating large numbers of people are daunting.
- Organizing rescue and recovery efforts in a stressed, multijurisdictional context is challenging.

Louisiana and Mississippi are not unique in their reliance on levees in flood risk management. Chronic underinvestment of resources in maintaining or upgrading critical levees has been noted by many for decades. For example, the important role of, and reliance on, levees was noted in the Midwest floods of 1993. Since 1998, the American Society of Civil Engineers has published a "Report Card for the Nation's Infrastructure," and although they did not "grade" levees separately until their 2009 report card (when levees received a "D-"), dams and navigational infrastructure consistently rated poorly with Ds and D-minuses.

3. If Levees Pose Such a Risk To Public Safety, why Hasn't Something Been Done Before?

There are many reasons why, individually and collectively, we have not acted to improve levee safety at a national level. Here are some factors:

- Part of our complacency is due to a misunderstanding of flood risk by decision makers and the general public. The National Flood Insurance Program uses the 1% annual chance flood (sometimes called the 100-year flood) as an actuarial standard. Homeowners living behind levees built to that standard are often exempted from the mandatory purchase requirement of flood insurance as a condition for a federally-backed mortgage, so many communities have built their levees to that standard and residents feel safe. The 1% annual chance level of flood protection is not a safety standard, however, and is frequently misinterpreted. Many believe this protection means that a destructive flood will only occur on average every 100 years. In reality, it means that there is a 1% chance every year that a flood of that magnitude will occur, translating to a 26% chance of flooding during a typical 30-year mortgage.
- As a nation, we do not understand – or even know – our risks. Without a national inventory, assessment, and inspection of all of the nation's levees, we do not know which ones protect the highest number of lives or the most critical infrastructure, or if those levees are providing the levels of protection expected.
- There is no single agency responsible for levee safety. Currently, responsibility for levee safety is dispersed throughout all levels of government and quasi-governmental agencies such as levee boards.
- There are no national standards for levee construction, inspection, or maintenance upon which to base our actions.

4. How Can I Keep My Family and My Home Safe?

Be informed. Know if you live, work, or go to school near or behind a levee. If you live near a major river or other body of water, there is a good chance that a levee may be nearby. Here are some ways you can check to see whether you live behind a levee:

- Contact a local government agency, such as your public works department or flood control district, to find out about whether levees are nearby and if your home depends on them to reduce the risk of flooding.
- Check www.floodsmart.gov and take the "One Step Flood Risk Profile" quiz. In it, you enter your address to learn if you live in an area at risk of flooding. If so, nearby levees may appear on the Flood Insurance Rate Map created by the U.S. Federal Emergency Management Agency (FEMA) to show the flood risk in your area. Follow the links on the site to FEMA's Map Service Center (msc.fema.gov) to download the map for your area. But be aware that levees are not always depicted on flood maps; your local public works department or flood control district may have more accurate information.

Be educated. There is always some risk living or working behind a levee. It is important to assess that risk. No levee provides full protection from flooding – even the best flood-control system or structure cannot completely eliminate the risk of flooding. Levees are designed to provide a specific level of protection, and larger flood events can cause them to be overtopped or fail. Levees also decay and deteriorate over time. Regular maintenance and periodic upgrades are needed to ensure that they retain their level of protection and continue to perform to their design. Maintenance can become a serious challenge as a levee system gets older. When levees do fail, they fail catastrophically – the damage may be more significant than if the levee wasn't present.

Your flood risk associated with a levee depends on two major factors: (a) the hazard or probability that you will be flooded, and (b) the potential loss of your property, the potential loss of your livelihood, or even the loss of your life or the lives of your loved ones as a result of flooding. This may seem complicated but it's really common sense. People with the highest risk: (a) live in floodprone areas with a high probability of being flooded, and (b) have property that, if flooded, would be expensive or impossible to replace. Note that your risk may change over time if risk factors change.

Be prepared:

- Understand the potential depth of flooding and time it will take to evacuate your home, if necessary. + Develop a personal emergency evacuation plan for yourself and your family.
- Keep important documents and duplicate records in a separate, safe place that is not prone to flooding.
- Purchase flood insurance – flood insurance could be an important element for safeguarding the financial investment in your home, even if you are not shown as being in a high flood risk area.
- Learn who is responsible for notifying residents of flooding in your area so you know who to turn to for information.
- When under a flood warning, locate your pre-assembled emergency kit (which should include a weather radio) and prepare other items to take with you in the event of an evacuation. Move valuable items from basements and ground-floor levels to higher areas. Turn off electricity at your breaker or fuse box and close your main gas valve. For fuel oil or propane tanks, turn off the fuel valve at the tank. Bring outdoor possessions inside or secure them adequately. Place sandbags anywhere water may enter your home. And, of course, if instructed to do so, leave immediately, avoiding areas of high or moving water.

Get involved. Encourage community officials to reduce the community's flood risk through a variety of approaches, including: converting areas in floodprone areas into parks or greenways that can accommodate flooding; remodeling or raising buildings to above floodwater levels; urging community homeowners to purchase flood insurance; avoiding building structures, planting trees, or leaving debris on a levee; changing zoning ordinances or building codes to limit development in floodplains; developing or refining flood warning systems, emergency evacuation plans, and community flood preparedness; and providing technical and/or financial assistance to property owners to protect against flooding.

(Adapted from So, You Live Behind a Levee!, American Society of Civil Engineers, 2009. Available online at http://content.asce.org/files/pdf /SoYou LiveBehindLevee.pdf)

The National Committee on Levee Safety

5. What is the National Committee on Levee Safety? Who is on the NCLS?

The NCLS was created by Congress and authorized by the Water Resources Development Act of 2007, Section IX, to "develop recommendations for a national levee safety program, including a strategic plan for implementation of the program." The NCLS includes representatives of state governments, local/regional governments, the private sector and two federal agencies (U.S. Army Corps of Engineers and the Federal Emergency Management Agency). A designee for the Assistant Secretary of the Army (Civil Works) chairs the NCLS.

The NCLS adopted the vision of "an involved public and reliable levee systems working as part of an integrated approach to protect people and property from floods." In their report entitled Recommendations for a National Levee Safety Program: A Report to Congress from the National Committee on Levee Safety (January 2009), the NCLS presented 20 recommendations that, when taken together, will establish the basis for a comprehensive and effective National Levee Safety Program.

6. What Is the Status of the NCLS's Report and Recommendations?

The NCLS presented their findings and recommendations for a National Levee Safety Program in a draft Report to Congress, Recommendations for a National Levee Safety Program: A Report to Congress from the National Committee on Levee Safety, completed in January 2009. The House Subcommittee on Water Resources and the Environment held a hearing on the recommendations in May 2009. The Office of Management and Budget also has reviewed the report.

The NCLS continues to work to provide further detail on some of its recommendations so that Congress can be as well-positioned as possible to create a National Levee Safety Program. As with all efforts that involve complex policy questions, extensive intergovernmental coordination, and significant technical standards development, the NCLS continues to work to identify and communicate to Congress which recommendations need legislation and which recommendations can begin moving towards implementation. The NCLS also intends to solicit additional feedback from both technical experts and stakeholders. The core of the recommendations made in the January 2009 Report to Congress, and the reasoning behind them, remain unchanged.

7. How Does the Report Reflect the Feedback the NCLS Received? How Did the NCLS Involve Stakeholders in Developing their Recommendations?

At the same time that the NCLS was created, a review committee representing interests from federal, state, local, and tribal agencies, interest groups, and technical societies also was created. During the three-month report development process, the NCLS held two review committee meetings to solicit feedback on progress to date. The first review committee meeting focused on the scoping of the report and whether the NCLS was asking the right questions. The second review committee meeting sought comment on high-level recommendations. The NCLS will continue to periodically seek advice from the review committee. The NCLS presented high-level draft recommendations and received comments at a webinar (virtual meeting), attended by about 300 people. The NCLS is now working to seek reaction and input to its recommendations to a broader set of stakeholders by conducting regional workshops as well as targeted outreach toward key implementers such as state, tribes, local governments, and levee operators.

8. What Is the NCLS Doing Now? The NCLS Is:

1. Working to further detail some recommendations to best position Congress to act to create a National Levee Safety Program, including:
    - Examining governance options for a National Levee Safety Program.
    - Further detailing benefits and costs of a National Levee Safety Program.
    - Expanding the NCLS's understanding of the complexities of the state of the law and jurisprudence relating to levee liability for local communities, states, and the private sector.

2. Supporting U.S. Army Corps of Engineers (USACE) and Federal Emergency Management Agency (FEMA) efforts to implement actions recommended in the NCLS's Report to Congress (January 2009) with existing funding and authorities, such as:
    - Expanding the USACE National Levee Database to include voluntarily submitted data from states to improve the understanding of the number, location, and condition of the nation's levees, necessary to fully understand the nation's risk. The NCLS's recommendations call for a onetime, federally

funded inventory and inspection of all the nation's levees, with that data incorporated in the National Levee Database.

- Further developing the important foundational aspects of several technical recommendations including: levee safety standards; a hazard potential classification system; and tolerable risk guidelines. It is the goal of the NCLS that by working with USACE to develop these standards and guidelines for federal levees, the groundwork will have been set for eventual creation and adoption of national standards. National standards require new authorities and funding.
- Working with FEMA and USACE to begin identifying and linking existing programs and websites related to levee safety, and to promote a more consistent use of terminology, messages, and approaches across the federal government.

3. Planning future outreach, including targeted efforts for states and tribes, to gain additional feedback on the design of participating levee safety programs.

9. Levee Safety Is Important; How Can I Help Establish a National Levee Safety Program?

Learn more about the recommendations for creating a comprehensive and effective National Levee Safety Program, and share that information with others. The full Report to Congress as well as shorter issue papers summarizing key issues and recommendations can all be found on the NCLS website,http://www.leveesafety.org.

- Attend a regional stakeholder workshop. The NCLS is conducting several workshops across the country to share information about the recommendations for a National Levee Safety Program and gather additional feedback, and you are invited. See the Events page on the website for details about upcoming events and to join the mailing list.
- If you support the recommendations or have questions or concerns, let us know and reach out to your local officials to share your support. The NCLS is working to lay the groundwork for the creation of the National Levee Safety Program, and ideas for improving levee safety nationwide are welcome. There is an area on the website to contact us with your feedback, questions, or concerns.

- If you are a member of a professional organization knowledgeable in levee safety issues, we would
- appreciate your engagement and support. Each element of the National Levee Safety Program is important, and you are welcome to reference and support the recommendations for a National Levee Safety Program in your own interface with stakeholders and decision makers. We welcome you to join us at upcoming regional stakeholder workshops (see the Events area of the website for more information).
- Understand your levee and your flood risk: Be informed. Be educated. Be prepared. Be involved.

10. Where Can I Find More Information on the NCLS and Its Recommendations?

The NCLS's website, http://www.leveesafety.org, includes a link to the report, issue papers on the NCLS and their recommendations, additional information on NCLS members, background documents, meeting minutes, and more.

The Recommended Elements of a National Levee Safety Program

11. What Did the NCLS Recommend in its Report to Congress?

First and foremost, the NCLS's recommendations for a National Levee Safety Program have a primary goal of increasing public safety by reducing the risks of catastrophic flooding associated with levees.

There is no single activity that will solve this problem, which has been decades in the making. The fix will not be immediate; it took decades for the present situation to evolve to its critical state, and it will take us many years to significantly improve levee safety. It will require a combination of technical solutions, behavior changes, and policy/program alignment by owners/operators, individuals, and governments at all levels. A concerted effort to understand and share responsibility for levee safety is the crux of the NCLS's recommendations. The NCLS has made 20 specific recommendations that fall into three main categories:

- Comprehensive and Consistent National Leadership is needed and should be provided through the creation of a National Levee Safety Commission that will develop a National Levee Safety Program. The

Commission will be charged with understanding and communicating risks associated with levees, developing national levee safety standards, facilitating important research, and developing and overseeing participating state and tribal programs.

- Strong Levee Safety Programs in All States is the cornerstone of an effective National Levee Safety Program, following a consistent set of safety standards and protocols, and providing assistance and oversight to local community efforts and levee owners/operators.
- Alignment of Existing Federal Programs is crucial in order that investments in our nation's levees and programs with investment in leveed areas are effective. All federal programs should align with the same goals of risk reduction, developing resilient and reliable levees, and protecting human life and property.

12. What Are the Benefits of a National Levee Safety Program? How Much Will a National Levee Safety Program Cost?

In their report, the NCLS concluded that the benefits of a National Levee Safety Program are substantial. Reducing the potential for loss of life and human suffering alone warrants the cost of the program. The recommendations also are a good financial investment by significantly reducing flood damages to the public and private sectors and reducing overall disaster relief costs. Based on data from the U.S. Army Corps of Engineers, the State of California, and the National Flood Insurance Program, it is reasonable to estimate that damages from levee failure or breaches reach more than $5 billion annually.

The NCLS made preliminary estimates of costs for all recommendations within the Report to Congress. They estimated that to implement its recommendations for a comprehensive and effective National Levee Safety Program, the federal share would be $878 million annually for the first five years of the program, and $360 million dollars annually in nonfederal cost share. The bulk of that federal investment ($600 million) would be in the proposed Levee Rehabilitation, Improvement, and Flood Mitigation Fund. In evaluating these estimates, one must remember that we do not yet know the magnitude of the problem because we do not have a complete inventory and inspection of the nation's levees.

13. Why Did the NCLS Recommend a New Bureaucracy/Agency? Why Not House the National Levee Safety Program in Either USACE or FEMA?

Currently, responsibility for levee safety is assigned in an often uncoordinated and incomplete manner, distributed across all levels of government, and housed in different agencies. Aligning federal agencies and working with states and local governments will require a level of independence from any one agency to be effective. In the report, the NCLS highlighted the following three guiding principles as essential and believes an independent federal agency with strong guidance by state, tribal, and local governments and the private sector is the ideal model:

(a) Independence to address levee safety holistically, unconstrained by the momentum and priorities of existing programs, with the ability to make politically challenging decisions when necessary.

(b) Leadership for significant horizontal integration of effort across federal agencies and alignment of their programs, with vertical integration to achieve balanced participation at all levels of government.

(c) Organizational capabilities spanning regulatory policy development, program implementation, oversight, and grants management, and significant experience in the relevant technical, public communication, and environmental areas.

While the NCLS strongly believes that an independent agency is preferable, they did consider the option of embedding the National Levee Safety Program in a single existing federal agency, such as the US Army Corps of Engineers (USACE) or the Federal Emergency Management Agency (FEMA). The NCLS determined that no existing agency alone has the full suite of expertise needed in the key areas of: 1) levee engineering; 2) risk mitigation in leveed areas; and 3) incentive and financial administration. The NCLS was also concerned that housing a national program in an existing agency would further stretch the resources of these agencies and would pose challenges to the chosen agency in coordinating alignment of other federal programs that they would be unlikely to overcome.

14. Why Did the NCLS Recommend Including Canals in the National Levee Safety Program?

The NCLS strongly believes that structures along canals "that constrain water flows and are subject to more frequent water loadings but that do not constitute a barrier across a watercourse" should be included in the National Levee Safety Program, as they pose many of the same risks as levees and can have a direct impact on public safety.

15. Why Is an Inventory of All the Nation's Levees Needed?

A national inventory of all levees is the first step in helping us understand the extent of the problem. Currently, there is no complete inventory of all the nation's levees. We know from an inventory of levees participating in U.S. Army Corps of Engineers (USACE) programs that there are over 2,000 USACE levee systems, totaling over 14,000 miles of levees – but the NCLS has estimated that there may be as many as 100,000 miles of levees across the nation based on information from USACE, the State of California, and the Federal Emergency Management Agency. It is important that we know what levees are in the public and private inventory. We then can assess the reliability of these levees, characteristics of the population and property in the leveed areas, and prioritize where funds need to be spent first to protect human lives and critical infrastructure.

16. Aren't there Standards for Levee Construction or Maintenance?

There are no national engineering standards for levee design, construction, operations, or maintenance; and various federal, state, and local agencies use different criteria. For example, the USACE has engineering policies, procedures, standards, and criteria for levees, but those standards only apply to levees that are enrolled in its programs; the U.S. Bureau of Reclamation maintains its own set of standards related to levees in its programs.

17. How Is the NCLS Addressing Concerns about Environmental Protection and Permitting?

One of the NCLS's recommendations is to develop and implement measures to more closely harmonize levee safety activities with environmental restoration and protection requirements. Some of the questions the Committee believes are essential to consider include:

- Can operations and maintenance practices necessary to maintaining the levee be improved to mitigate or reduce negative impacts on the natural environment or ecosystem without compromising public safety?

- Can levees or flood risk mitigation projects, including operations and maintenance activities, be designed or modified to reduce negative environmental impacts – or even to enhance or restore the environment?
- Can streamlining permitting requirements related to levee operations, maintenance, and emergency repairs, and reducing the time required to obtain those permits, improve both the natural environment and public safety, increasing efficiency without shortcutting either the environment or public safety?

The NCLS is recommending examining opportunities to ensure that levee operations can be conducted in a manner respectful of the environment and protecting public safety. They have recommended that a standing Advisory Committee on Environment & Safety be established to provide advice on the coordination of environmental and safety concerns related to levee operations and maintenance, removal, rehabilitation, and new levee projects.

18. Why Is the NCLS Concerned about Liability?

Anecdotal evidence suggests that liability concerns in government and industry may impede development and implementation of strong state levee safety programs throughout the nation. Concern also is growing in the engineering community and among levee districts, owners, and operators regarding potential liability that may be incurred through performance of levee services. These services include certification of data for National Flood Insurance Program accreditation purposes, as well as levee design, construction, and maintenance services. Informal surveys suggest that at least some engineering firms are declining to perform levee work, in part out of liability concerns. The NCLS is encouraging research to identify the current state of the law as it applies to liability. Research should also explore possible measures to prevent liability concerns from hindering development of robust levee safety programs.

19. Why Are State Programs Recommended as Part of the Federal National Levee Safety Program?

Strong levee safety programs at the state and local level are a crucial component of an effective national program. The NCLS believes that states are best positioned to organize, implement, and oversee levee safety programs, as they have the combination of necessary legal authorities, statewide reach, and existing relationships with local governments to be successful. It is the intent of the NCLS's recommendations that states could allow for some regional and state variation and tailoring in programs to meet local needs and conditions while meeting national standards and objectives, making the effort more effective. Finally, states already are playing similar roles with regard to other infrastructure programs such as dam safety, water and wastewater treatment, etc.

Further, the NCLS understands that states cannot implement this program alone, nor can it be an unfunded mandate, and so designed an approach that encourages states by providing start-up grant funding for states to develop levee safety programs, as well as technical assistance to help develop the myriad of technical expertise needed (e.g., engineering specialization, risk communication, etc.) at all levels of the state. As the state programs mature, it is anticipated that further incentives, through preference in other federal programs, will be afforded to states that implement strong levee safety programs, and conversely disincentives established for states that take little or no action to address their shared responsibility for levee safety.

State representatives on the NCLS have played a key role in developing these recommendations. The NCLS will involve the states more heavily in the specific design of participating levee safety programs through expanded stakeholder involvement efforts.

20. Why Did the NCLS Recommend Risk-Based Flood Insurance in Leveed Areas?

Flood insurance is one of the most effective ways to limit financial damages in the case of flooding and speed recovery of flood-damaged communities. However, despite the fact that the 1% annual chance flood standard used by the National Flood Insurance Program is not a safety standard, the exemption from flood insurance requirements have led many individuals and communities in leveed areas to mistakenly believe that they do not need flood insurance, and that they are protected from all flooding by that levee.

The NCLS has recommended that every structure in a leveed area should have flood insurance, and that the insurance premiums should be risk-based. That is, the insurance premiums would be developed to reflect the flood risks of living behind a levee and considering the levee's level of flood protection. The NCLS's recommendations aim at increasing the understanding that living behind even well-engineered levees has some risk. Because the recommendation is for risk-based premiums, it will help incentivize local communities to maintain reliable levees, good evacuation programs, flood proofing, etc., as a way to lower premiums for their constituents in leveed areas.

Without some significant regulatory changes by the Federal Emergency Management Agency (FEMA), existing mandatory flood insurance purchase requirements for the National Flood Insurance Program (NFIP) mandated by the Flood Disaster Protection Act of 1973 and National Flood Insurance Reform Act of 1994 would not achieve the Committee's goal of risk-based flood insurance for all structures in leveed areas. However, the NFIP Reform process currently underway provides an opportunity for FEMA to implement this recommendation, and the Committee plans to be engaged throughout the process.

Depending on the outcomes of the NFIP Reform process, additional mechanisms may necessary in order to meet the objective of flood insurance for all structures in leveed areas, although the Committee has not explored other federal, nonfederal, or private sector approaches to promote or require flood insurance in leveed areas.

It is important to note that no levee provides full protection from flooding – even the best flood-control system or structure cannot completely eliminate the risk of flooding. Levees are designed to provide a specific level of protection, and larger flood events can cause them to be overtopped or fail. Levees also decay and deteriorate over time. Regular maintenance and periodic upgrades are needed to ensure that they retain their level of protection and continue to perform to their design. Maintenance can become a serious challenge as a levee system gets older. When levees do fail, they fail catastrophically – the damage may be more significant than if the levee wasn't present.

21. Why Did the NCLS Not Go Further into More Floodplain Management Recommendations?

The NCLS's recommendations are prefaced by recognition of the need for a broader national flood risk management approach. The legislation creating the NCLS limited its charge to developing recommendations for a National Levee Safety Program. It is hoped that by opening this dialogue on levee safety, the larger issue of floodplain management also will be considered and addressed.

Relationship to other Federal Efforts

22. Doesn't USACE Already Have a Levee Safety Program?

The U.S. Army Corps of Engineers (USACE) does have a Levee Safety Program, but it does not have either the authority or the expertise to implement the comprehensive National Levee Safety Program recommended by the NCLS. The USACE Levee Safety Program, established in 1996, conducts inspections and assessments of approximately 14,000 miles of levee systems; evaluates, prioritizes, and justifies levee safety decisions; incorporates changes and improvements associated with state-of-the-art professional engineering practice into its levee safety policy and procedures; and makes recommendations to improve public safety associated with levee systems. The authorities of the USACE Levee Safety Program are generally limited to levees operated and maintained by USACE, federally authorized projects in the USACE Inspection of Completed Works program, and nonfederally authorized projects within the USACE Rehabilitation and Inspection Program.

23. What Is the Difference between the Levee Inventory USACE is Conducting under WRDA 2007 and the NCLS Recommendation for a Levee Inventory?

The NCLS recommendations call for a one-time, federally funded inventory and inspection of all the nation's levees, with that data to be incorporated in the National Levee Database.

The levee inventory being undertaken by the U.S. Army Corps of Engineers (USACE), authorized by WRDA 2007 (the Water Resources Development Act), is of levees participating in the USACE Levee Safety Program, that is: levees operated and maintained by USACE; federally authorized projects in the USACE Inspection of Completed Works program; and nonfederally authorized projects within the USACE Rehabilitation and Inspection Program. This is an important first step in that many of these levees are large levees, with millions of people, property, and critical infrastructure behind them. However, it only includes approximately 14,000 miles of the estimated 100,000 miles of levees believed to exist in the nation.

Initially populated with data from its levee inventory, USACE is expanding the National Levee Database to include voluntarily submitted data from states in order to improve our understanding of the number, location, and condition of the nation's levees, a necessary step to fully understand our nation's risk.

24. If I Live Behind a Levee Accredited in the National Flood Insurance Program, I'm Not at Risk of Flooding. Right?

One of the biggest misunderstandings related to levees is that accreditation under the National Flood Insurance Program (NFIP) means that you are not at risk from flooding. In 1968, Congress enacted the NFIP with the primary purpose of addressing the inability of the public to secure privately backed insurance for economic losses due to flooding. The NFIP uses the protection level of a 1%-annual-chance flood (sometimes called the 100-year flood) as an actuarial standard. Homeowners living behind levees built to that standard are often exempted from the mandatory purchase requirements for flood insurance as a condition for a federally-backed mortgage, so many communities have built their levees to that standard and residents feel safe. The 1% annual chance level of flood protection is not a safety standard, however, and is frequently misinterpreted. Many believe this protection means that a destructive flood will occur on average only every 100 years. In reality, it means that there is a 1% chance every year that a flood of that magnitude will occur, translating to a 26% chance of flooding during a typical 30-year mortgage.

Unfortunately, the 1% annual chance, never intended to be a safety standard, became a design criterion for many communities and unrestricted development was allowed in leveed areas, greatly increasing the numbers of people and amount of property at risk from flooding.

Further, accreditation of a levee under the NFIP does not guarantee that the levee will not fail or overtop. Please consider: 1) storms and flooding exceeding the 1% annual chance can and do happen on a regular basis, additional development in the floodplains and stronger storms due to a changing climate further increase the chances of higher and more violent flooding; 2) many levees, even those well maintained, are aging and have not been updated to meet modern engineering standards; and 3) all levees can fail.

Bottom Line: The accreditation of levees under the National Flood Insurance Program merely ensures that the levee has met the technical standards for providing an expected level of flood protection. It does not guarantee that the levee will not be overtopped – and your home flooded –

during a larger storm event, or that the levee cannot be damaged or breached unexpectedly. While living behind an accredited levee reduces your risk of flooding, it does not eliminate it. Flood insurance is one of the most effective ways to limit financial damages in the case of flooding, and speed the recovery of flood-damaged communities.

25. How Do the Recommendations of the National Levee Safety Commission Relate to the Updating of the Principles and Guidelines?

Overall, the Administration's effort to update the Principles and Guidelines for Water and Land Related Resources Implementation Studies (known as "The Principles and Guidelines" or "P&G") – with its stated goal of better understanding costs and benefits of federal actions in a broader fashion, including social, environmental and public safety impacts, as well as economic benefits – is consistent with the recommendations of the NCLS for a National Levee Safety Program.

Levees, Levee Safety, and Risk

26. My Levee Looks Good and Is Regularly Maintained. Why Do You Think I Am at Risk?

Levees are designed to provide a specific level of protection, and larger flood events can cause them to be overtopped or fail; we simply can't afford to build levees to keep all storms out. Even levees in good repair are subject to overtopping in storms of great magnitude. Further, well-maintained levees may have been built without adequate engineering and may fail before they are overtopped. It is also important to remember that the state of the art of engineering has evolved. If your levee is dated, it may not be up to today's standards.

Because the risk of property damage and loss of life still remains, you should take steps to lessen that risk through considering such actions as purchasing flood insurance, raising the elevation of your home, and having and implementing an emergency evacuation plan when the need arises.

In: National Levee Safety…
Editors: D. Mancuso, V. Masterson
ISBN: 978-1-61470-511-6

Chapter 3

LOCALLY OPERATED LEVEES: ISSUES AND FEDERAL PROGRAMS*

Natalie Keegan, Rawle O. King, Nicole T. Carter and Megan Stubbs

SUMMARY

Locally operated levees and the federal programs that assist and accredit them are receiving increasing congressional attention. Congressional authorization of the National Flood Insurance Program (NFIP), managed by the Federal Emergency Management Agency (FEMA), expires on September 30, 2011. The pending reauthorization has increased congressional awareness of the link between the condition of locally operated levees, FEMA's Flood Insurance Rate Maps (FIRMs) and levee accreditation (which determine which NFIP requirements and premiums apply in an area), and programs providing federal disaster assistance for these levees. Congress is considering whether and how to change current programs, federal funding, and the existing division of levee responsibilities. Options are complicated by the desire to promote state, local, and individual decisions and investments that reduce flood risk; concerns about the local costs associated with NFIP purchase and levee accreditation requirements; and consideration of whether to expand federal responsibilities and potential liability. Even though similar issues

* This is an edited, reformatted and augmented version of Congressional Research Service Report R41752, dated April 5, 2011.

also exist for some of the federally operated levees, this report focuses on locally operated levees since these dominate the national levee portfolio.

Approximately 22% of U.S. counties across the country, representing almost half of the U.S. population, contain levees. Economic damage from floods in leveed areas ranges between $5 billion and $10 billion annually. Levees play an important role in protecting property against flood damage. More than 100,000 miles of levees may exist, with the federal government operating roughly 2,100 miles. One estimate puts the five-year level of investment needed for new construction or maintenance of the nation's levees at $50 billion.

FEMA is updating FIRMs and deciding whether to accredit levees which will determine whether they appear on those maps. There is some debate regarding the extent to which FEMA should assist with levees investment through its hazard mitigation programs. FEMA often cites overlap with activities of the U.S. Army Corps of Engineers (Corps) and the Natural Resources Conservation Services (NRCS) in the U.S. Department of Agriculture as justification for not funding levee activities.

The Corps is the main federal partner for construction of locally operated levees. Pursuant to congressional authorizations, the Corps participates in cost-shared planning and construction of levees. No general federal authorities exist for the Corps to assist with the regular operation and maintenance of locally operated levees; that is, local levee owners are responsible for operation, maintenance, and improvement. However, there are multiple authorities enacted by Congress for flood fighting, flood mitigation, and levee repair of damages caused by natural events. Since 2005, the Corps has had limited involvement in the data collected and certified to inform FEMA accreditation of locally operated levees. The Corps has limited authority to assist local levee owners in obtaining NFIP levee accreditation. Policymakers in recent years have considered whether to expand the Corps' role in NFIP data certification and post-construction improvements for locally operated levees. NRCS has limited authority to assist in the construction of smaller levees and to repair small, mostly rural levees damaged by a natural event.

Congressional options for assisting with levees include, but are not limited to, maintaining the status quo, adopting the recommendations of the National Committee on Levee Safety (such as federal support to develop new state levee safety programs), modifying federal programs, or creating new federal programs.

Introduction

Congressional awareness of issues associated with locally owned and operated levees is increasing, largely because of the nationwide remapping of floodplains by the Federal Emergency Management Agency (FEMA) and efforts to reauthorize the FEMA's National Flood Insurance Program (NFIP) which expires on September 30, 2011.[1] Floodplain remapping can result in owners of mortgaged structures located in areas protected by levees being required to purchase flood insurance if the levees cannot satisfy FEMA's levee accreditation requirements. Many communities and levee owners struggle with the expense and effort of data collection, repairs, and improvements needed to obtain levee accreditation and have approached Congress for assistance.[2] Multiple federal agencies have roles in levee accreditation and federal levee assistance. Most notably, FEMA coordinates updates to Flood Insurance Rate Maps (FIRM) and issues levee accreditations that waive the mandatory purchase requirement for homeowners who reside in areas protected by levees. In certain limited situations discussed later in this report, the U.S. Army Corps of Engineers (Corps) may certify the data submitted to FEMA for accreditation purposes. However, existing federal actions to assist locally operated levees are limited, particularly in the scope of eligible activities. Limitations stem from constraints on federal funding and from levee operation, maintenance, and improvements generally being the responsibility of the levee owner.[3]

Levees are structures, either earthen embankments or concrete and steel floodwalls, built along rivers, or other bodies of water, to prevent water from flooding bordering land.[4] Levees are located throughout the country and are found in approximately 22% of U.S. counties, where almost half of the U.S. population resides. Nationally, average economic damage from floods in leveed areas ranges between $5 billion and $10 billion annually. The full extent and specific conditions of the nation's levees are unknown, but the National Committee on Levee Safety, which was created by Congress, estimates that more than 100,000 miles of levees may exist.[5] The federal government through the Corps built less than 15,000 miles of these levees.[6] The Corps operates roughly 2,100 miles of the 15,000 it constructed. That is, local levee owners operate and maintain the estimated 85,000 miles of locally constructed levees and almost 13,000 miles of Corps constructed levees. One estimate puts the five-year level of investment for new construction or maintenance needed for the nation's existing levees at $50 billion.[7]

Congress has considered in recent years whether and how to change the current division of levee responsibilities and their costs, and whether to modify existing levee-related federal programs. Some stakeholders seek to expand flood hazard mitigation activities eligible for federal funds to include levees, while others are concerned that the federal government might assume more of the costs and liability for levee investments that they consider a local responsibility. Another aspect of the debate is whether to change FEMA's risk assessment, remapping, and timelines for obtaining levee accreditation. These changes are raised in the context of climate change and a broader rethinking of flood risk management and control structures (like levees) following Hurricane Katrina and recent floods.

This report covers the federal role in locally operated levees. It does not address federally owned and operated levees, which are less common and are concentrated in a few areas of the United States (e.g., along the Mississippi River).

The report first discusses the role of levees in flood risk reduction, the shared responsibilities for levees in the United States, and the role of three agencies: FEMA, the Corps, and the Natural Resources Conservation Service (NRCS) of the U.S. Department of Agriculture (USDA). Next, it discusses federal assistance for levees starting with their study and construction, then levee-related flood fighting and levee repair, rehabilitation, and inspection. It also describes the debate about whether levees investments have a role in federal flood mitigation programs. Corps, FEMA, and NRCS activities and authorities are compared. Finally, the report outlines policy options for locally operated levees that might be considered by the 112th Congress. Legislative proposals in the 111th Congress are discussed in an *Appendix*.

FLOOD RISK REDUCTION AND LEVEES IN A FEDERAL SYSTEM

In the United States, flood-related roles and responsibilities are distributed within a regulatory framework designed to permit the responsible development of the nation's floodplains. Local governments generally are responsible for land use and zoning decisions that shape floodplain and coastal development, but the federal and state governments also influence decisions on managing flood risk. State and local governments make decisions that allow or prohibit development in flood-prone areas. Local, private, and sometimes state entities construct, operate, and maintain most levees and have initial flood-fighting responsibilities. Agencies of the federal government operate flood control dams, offer crop insurance, support hazard mitigation, and provide emergency response and disaster aid for recovery from floods. The federal government constructs some of the nation's levees, but most often does in partnership with local project sponsors who are responsible for operation and maintenance.

No federal program specifically regulates the design, placement, construction, or maintenance of nonfederal levees. While federal and state agencies often have their own policies and criteria for levee design, construction, and maintenance, there are no national engineering standards[8] or policies for levee design or level-of-protection standards for different floodplain uses.

Many levees protecting today's communities and agricultural investments originally were constructed more than 50 years ago by local interests reclaiming land for agriculture and other uses. Rather than each landowner building separate levees, landowners often consolidated their resources by forming a levee district. As a consequence of this history, many of today's physical constructions and configurations, as well as institutional arrangements for flood protection, have roots distinct from their current use as flood protection for development. For the most part, municipalities serving concentrated urban populations have assumed flood control responsibilities, while special levee districts remain active in rural and agricultural areas.

Prior to the Lower Mississippi River Flood of 1927, the federal role in flood control was limited. In addition to authorizing the Corps to design and construct significant flood control projects along the Mississippi River (and on the Sacramento River (CA)), the Flood Control Act of 1928 reiterated the sense of Congress, at the insistence of President Coolidge, that there should be local contribution toward flood control works.[9] The act also provided that the federal government generally could not be held liable for flood damage. From 1928 to 1936, there was considerable debate about the need for national planning for flood control and a national water resources program, and the proper roles of Congress and the Executive Branch in this planning and selection of construction projects.[10] The Flood Control Act of 1936 (49 Stat. 1570) declared some flood control a "proper" federal activity:

> It is hereby recognized that destructive floods upon the rivers of the United States, upsetting orderly processes and causing loss of life and property, including the erosion of lands and impairing and obstructing navigation, highways, railroads, and other channels of commerce between the

> States, constitute a menace to national welfare; that it is the sense of Congress that flood control on navigational waters or their tributaries is a proper activity of the Federal Government in cooperation with States, their political sub-divisions and localities thereof; that investigations and improvements of rivers and other waterways, including watersheds thereof, for flood-control purposes are in the interest of the general welfare; that the Federal Government should improve or participate in the improvement of navigable waters or their tributaries including watersheds thereof, for flood-control purposes if the benefits to whomsoever they may accrue are in excess of the estimated costs, and if the lives and social security of people are otherwise adversely affected.

Since 1936, Congress has authorized the Corps to construct hundreds of miles of levees.

Today the federal role in flood risk management goes beyond assisting with the construction of dams and levees. For instance, some federal actions attempt to modify individual and community behavior to reduce flood vulnerability; the NFIP and FEMA's hazard mitigation programs are examples. Congress established the NFIP with the 1968 passage of the National Flood Insurance Act.[11] Prior to the program's establishment the federal government relied on a "levee-only" policy that permitted unrestricted development of the floodplains along with an emphasis on structural flood control systems. Making federally subsidized flood insurance available to property owners signaled a shift in federal policy towards a flood control strategy that was less dependent on structural measures. The new policy had the intended consequence of regulating the development of flood-prone areas. The NFIP's multi-pronged regulatory system consists of flood risk assessment and mapping, flood insurance, and land use and building construction measures that restricted development in vulnerable areas. The NFIP allows for residential construction in known floodplains, with the proviso that construction must follow building code regulations that include flood-proofing requirements. FEMA's hazard mitigation programs fund mitigation activities such as elevating properties, acquiring properties and converting them to open space, retrofitting buildings, and implementing limited flood control systems.

Divided Federal Responsibilities

Federal agencies play various roles in planning, construction, maintenance and operation, repair and rehabilitation of levees and related flood insurance and mapping.[12] Three agencies are authorized to provide federal assistance for locally owned or maintained levees—FEMA, Corps, and NRCS. FEMA has responsibility for NFIP flood risk assessment, mapping, and levee accreditation. The Corps performs some levee construction and damage repair and has the largest federal appropriations for these activities. NRCS provides some funds for repair of damaged levees through its Small Watershed Program. The levee-related roles of the three agencies are discussed below and in *Table 1.*

Federal Emergency Management Agency

FEMA is responsible for flood risk assessment and mapping, flood insurance, and federal hazard mitigation. Much of the congressional attention concerning locally operated levees derives from concerns about insurance expenses under the NFIP. FEMA uses FIRMs to delineate flood risk zones and the applicable insurance premiums to be charged for properties covered by federally backed mortgages.

Table 1. FEMA, Corps, and NRCS Authorities for Locally Operated Levee Work

Federal Agency	Study and Construction	Operation and Mainte-nance	Emergency Flood Fighting	Repair of Damage	Improve ments	NFIP Data Certifi-cation
FEMA	Authority uncleara	No authority	Stafford Act (42 U.S.C. 5170a)	Authority uncleara	Authority uncleara	No authority
Corps	Congressionally authorized actions(33 U.S.C 701n)	No authority	Emergency response authority(33 U.S.C. 701n)	Rehabilitation and Inspection Program(33 U.S.C. 701n)	No authority	Limited authority
NRCS	Small Watershed Program (33 U.S.C. 701b-1) and (16 U.S.C. 1001. et seq.)	No authority	No authority	Emergency Watershed Program(33 U.S.C. 701b-1) and (16 U.S.C. 2203)	No authority	No authority

Source: CRS analysis of FEMA, Corps, and NRCS programs.

a. 42 U.S.C. 4104c provides FEMA with the authority to undertake flood mitigation activities. It is unclear to what extent such authority would extend to locally operated levees within the regulatory constraints of the prohibition on duplication of federal programs, including Corps and NRCS programs.

FEMA is updating the nation's inventory of FIRMs and requiring verification, through accreditation, that all levees currently depicted on FIRMs meet design, operation, and maintenance standards for protection against the 1%-annual-chance flood. FEMA activities have generated a number of questions. Some communities and stakeholders have raised questions about the development and accuracy of the updated FIRMs. Some levee owners have questioned the costs and documentation requirements associated with levee accreditation. Questions concerning liability in the event of a levee failure have been raised, most notably by representatives of engineering firms.

FEMA also operates various flood hazard mitigation grant programs that assist in removing repetitively flooded structures from floodplains, and conducts other activities to reduce flood impacts. Policy discussions surrounding these programs have questioned to what extent levee work should be eligible for FEMA mitigation funds. While some communities and levee owners want levee work to be funded by these programs, FEMA argues that funding levee projects would duplicate other federal programs and that appropriate provisions have not been made to address levee liability.[13]

Under FEMA regulations, hazard mitigation funds cannot replace project or program funding available under other federal authorities, unless there exists an extraordinary threat to lives, public health or safety, or improved property.[14] This provides FEMA officials with discretion to determine whether other federal programs are more appropriate to fund levee projects. Because certain levee activities receive funding from the Corps or NRCS, FEMA has determined that these agencies have the primary authority for the repair of flood control works such as levees.[15]

As a result, FEMA hazard mitigation assistance grants have not been used for levee projects for at least the last decade.16 While FEMA officials cite program duplication as justification for denying levee projects mitigation funding, it can be argued that FEMA has not consistently applied the same standard regarding duplication of authorities. That is, FEMA officials have allowed Hazard Mitigation Grant Program (HMGP) funds to be used for disaster housing projects that were eligible for funding under the Community Development Block Grant (CDBG) program, administered by the Department of Housing and Urban Development (HUD).[17] In fact, in some instances CDBG funds were used to augment HMGP-funded projects.

Army Corps of Engineers

The U.S. Army Corps of Engineers is responsible for much of the federal construction of flood control and storm protection infrastructure. At the direction of Congress, the Corps is authorized to participate in the cost-shared planning and construction of flood damage reduction projects, such as levees and floodwalls to reduce damages from riverine and coastal flood hazards. Appropriations for these Corps construction projects have not kept pace with authorizations, and there is a significant backlog of congressionally authorized studies and construction projects. Interest in expanding Corps levee responsibilities raises questions about how to prioritize the federal funds Congress appropriates for the Corps, given the competing demand for its water resources projects.

As shown in *Table 1,* the Corps has a limited role in the regular operation, maintenance, and improvement of locally operated levees. After construction, levees built by the Corps generally are turned over to a local entity for operation, maintenance, repair, and rehabilitation.[18] Congress has authorized the agency to fight flooding at locally operated levees during emergencies in order to protect life and improved property (i.e., not levees protecting agricultural lands). The Corps can repair damage caused by a natural event at a levee that participates in the Corps' Rehabilitation and Inspection Program (RIP, also known as the P.L. 84-99 program). The Corps currently has no general authority, responsibility, or funding to assist local levee owners in assembling their NFIP levee accreditation packages, unless the levee is part of an ongoing Corps study or project.[19] However, expanding the Corps' role in NFIP data certification and post-construction improvements of locally operated levees is being discussed as part of the policy debate on how to manage flood risk and promote risk reduction nationally.

Natural Resources Conservation Service

NRCS also funds levee projects, but on a smaller scale than the Corps. Most of these levees protect agricultural areas and typically do not provide a level of protection that can withstand large-scale flood events. Therefore, the NRCS role is often not raised in the context of the NFIP remapping and levee accreditation debates. NRCS programs, however, are part of the discussion regarding federal assistance for levees because FEMA often points to these programs as a rationale for denying levee projects under its hazard mitigation programs.

Levee Liability Issues

Liability risk for levee services (e.g., design, construction, maintenance, inspection, and data certification) may limit interest among public and private entities in providing such services and participating in levee projects. For public entities, this concern stems in part from the uncertainty related to the implications of Paterno v. State of California, which held the State of California liable for a levee it did not build, but operated as part of a state-sponsored levee system.20 Anecdotal evidence suggests that some engineering firms have declined to perform levee work because of liability concerns; however, other anecdotal information suggests that levee owners generally have found a firm or some other means to perform the needed work (e.g., city engineers signing NFIP levee data certifications). The additional cost to levee owners from efforts of private engineering firms to cover their potential liability for the levee services is unknown.

To some extent, federal agency liability for federally funded levee projects has been addressed through congressional action.21 One source of the federal government's immunity is the exception under the Federal Tort Claims Act for actions that constitute a discretionary function.22 A second source of immunity for the federal government is the Flood Control Act of 1928, which prevents the government from being sued for damages resulting from federally supported damage reduction projects or from floodwaters.23 However, as discussed in the later section "Policy Options for Assisting Locally Operated Levees," some questions remain as to whether the immunity applies to all federal actions and agencies.

National Flood Insurance Program

FEMA's NFIP plays a significant role in federal levee policy and promotes interaction between FEMA and the Corps on levee accreditation. As discussed earlier, the 1968 National Flood Insurance Act established the NFIP. Federal flood insurance currently is available in participating communities to help individuals and small businesses recover from flood damage. FEMA officials point out that the NFIP has realized significant savings both to itself and to property owners by reducing the cost of disaster relief. The basic requirement of the flood management program (and access to federal flood insurance) is that communities adopt and enforce minimum land use and building code regulations to prevent new development from increasing the flood risk and to protect new and existing buildings. Property owners obtaining loans from federally regulated lending institutions, or receiving federal assistance for acquisition or construction in special flood hazard areas (SFHAs) in communities that participate in the NFIP, are required to purchase flood insurance for their outstanding mortgage balance, up to a maximum of $250,000 in coverage for single-family homes. The mandatory purchase requirement applies only to certain properties in floodplains. Levees that protect a community from a 1%-annual-chance flood can reduce the NFIP mandatory purchase requirement.

Levees in NFIP Flood Risk Assessment and Mapping

FEMA is required to identify special flood, mudslide or flood-related erosion hazards within the community, establish appropriate risk zone determinations, and reflect these determinations accurately on FIRMs.[24] Flood maps generally delineate both high-risk zones and low-to-moderate risk zones (with a less than 1%-annual-chance of flooding). Flood maps have many uses, including local land-use planning, emergency preparedness and response, and natural resource management. Lending institutions and insurance companies also use FIRMs to calculate flood insurance rates and determine who is required to comply with mandatory purchase requirements. Therefore, how a levee is depicted on the FIRM has multiple consequences.

In 2003, at the urging of Congress and in collaboration with state and local partners, FEMA launched a five-year public awareness and map modernization program (Map Mod) to convert existing paper FIRMs into more accurate digital flood insurance rate maps (DFIRMs). As part of the Map Mod, FEMA initiated a nationwide flood insurance study (FIS) to update flood-hazard data used to identify the flood hazard risk in "levee-impacted" areas (i.e., areas protected by levees) on DFIRMs. According to FEMA, the primary goals of the FIS and DFIRMs are to:

- incorporate available flood hazard information;
- convert the base flood elevation data from the National Geodetic Vertical Datum of 1929 to the North American Vertical Datum of 1988; and
- upgrade the FIRMs to a geographic information system (GIS) database format.

As part of FEMA's Map Mod, FEMA also implemented a policy requiring verification that all levees currently depicted on FIRMs meet design, operation, and maintenance standards for protection against the 1%-annual-chance flood. FEMA reviews compliance with design, operation, and maintenance standards during the levee accreditation process.[25] This process has increased congressional interest in federal assistance for compiling the accreditation package materials and for investments in levee improvements to meet FEMA's accreditation requirements.

FEMA Levee Accreditation

As of November 2010, FEMA has accredited approximately 4,800 levees (which includes both locally and federally operated levees).[26] It is unclear what percentage of levees this represents since it is unknown how many levees exist nationwide. FEMA reviews levees for accreditation at the request of the entity that owns and operates the levee. The levee accreditation applications are reviewed as they are submitted. The regulatory requirements for accrediting levees as providing base flood protection are found in Title 44 of the *Code of Federal Regulations*, Section 65.10. On August 22, 2005, FEMA issued Procedure Memorandum No. 34 to clarify that while FEMA accredits the levees, it is the community or levee system owner's responsibility to provide data and documentation to show that a levee system meets the requirements of 44 C.F.R. § 65.10.[27] A levee accreditation allows development in the "levee-impacted" area to be exempt from the NFIP mandatory purchase requirements and land use restrictions.

Technical Criteria for Levee Accreditation

In its process for deciding whether to accredit a levee, FEMA does not conduct a detailed physical levee inspection to determine whether, and how, the levee will perform in a flood. Rather, officials from FEMA's regional office typically meet with county and local community officials to put together a levee accreditation plan outlining a process and timeline to submit a certification of the levee data to FEMA. The certified levee data must document that:

- the levee is designed to withstand forces from the 1%-annual-chance flood based on its height, stability, foundation stability, and embankment protection;

- the levee has adequate freeboard (e.g., the levee height is at least three feet above design flood state);[28]
- all closure devices function properly;
- operation and maintenance plans are adequate and in place; and
- interior drainage systems (pumps and canals) are functioning.

FEMA's accreditation is not a levee performance guarantee. It is only a statement that data and documentation submitted to the agency comply with FEMA guidelines.[29] If a levee that was previously accredited is found to no longer comply with FEMA standards, it is de-accredited.

Costs of Obtaining Levee Accreditation

Owners of locally operated levees are responsible for the costs associated with seeking and maintaining FEMA's levee accreditation. The costs of obtaining accreditation include producing the information necessary to certify that the levee provides protection from the 1%-annual-chance flood event, and making improvements to the levee to enhance its functioning so that it meets FEMA's accreditation requirements.[30] Data on how much levee owners are investing in levee improvements and in data certification in order to obtain FEMA accreditation are not available.

Some levee owners have expressed concerns about the costs and process for obtaining accreditation. The effectiveness and efficient functioning of FEMA's levee accreditation process have not been the subject of an independent assessment to date. Without an assessment, and its supporting data, it is difficult to identify whether, and how, the accreditation process could be improved.

Data Certification

FEMA requires that a professional engineer certify that a levee complies with all requirements established in 44 C.F.R. § 65.10(b) before it can appear on a FIRM; this process is known as levee data certification. The levee data certification package is used by FEMA as the basis of its determination for whether or not to accredit a levee.

In 2005, FEMA increased the information requirements needed to accredit a levee. FEMA now requires more data on the structural integrity of the levee and the hydrology and hydraulics to which the levee is exposed than in the past.[31] As a result, local owners of some levees that previously were accredited by FEMA are having trouble obtaining and paying for reaccreditation. In many cases, they face a lack of readily available data on their levees' construction, materials, and structural integrity and confront assessments that indicate a lower level of protection than previously thought.

To complete the levee data certification, a professional engineer must sign a statement that the levee is designed in accordance with sound engineering practices to provide protection from the 1%-annual-chance event. Defining what constitutes sound engineering practices is complicated by the absence of a national levee model design for engineers to turn to as the professional standard for levee design parameters. Some engineering firms and their insurers have been concerned about liability if a firm certifies levee data and the levee later fails. Proposals for addressing this concern include providing immunity for the firms and clarifying that the professional engineer is certifying that the data comply with FEMA's requirements, rather than attesting to the levee's safety. Whether a federal agency certifying levee data would be immune from similar liability concerns is unclear and may be an issue if Congress considers increased Corps participation in data certification.

Prior to the 2005 increased information requirements, FEMA accepted the Corps' inspection of levees that participated in the Corps' Rehabilitation and Inspection Program (RIP) as sufficient for the data certification for levee accreditation. Since 2005, Corps RIP inspections are insufficient to meet the current FEMA data certification requirements.[32] The purpose of RIP inspections is to assess compliance with the requirements of RIP — that the levee owner is performing the maintenance required for RIP participation. The RIP inspections do not evaluate a levee's level of protection and structural integrity as required by FEMA for levee accreditation. In August 2010, the Corps released new guidance on how the agency will conduct the NFIP levee data certifications (which it calls levee system evaluations) that it has the authority to perform (see box "Limited Corps Role in Levee Data Certification").[33] The Corps process complies with the FEMA NFIP requirements under 44 C.F.R. § 65.10(b), but uses a different approach. The Corps approach is more focused on an evaluation of flood risk with the levee in place, while the 44 C.F.R. § 65.10(b) requirements are more focused on the level of protection provided by the levee.

Limited Corps Role in Levee Data Certification

Some levee owners have looked to the Corps to assist with levee data certification. The Corps does perform data certification for federally operated levees, for locally operated levees that are part of a larger ongoing Corps study or project, and at the request of another federal agency.

The Corps currently has no general authority to perform NFIP-compliant data certifications using discretionary appropriations for locally operated levees and is restricted from performing FEMA data certification on a reimbursable basis for nonfederal entities if the work can be provided by the private sector. This restriction is established for all Corps civil works activities of §211 of the Water Resources Development Act (WRDA) of 2000 (P.L. 106-541, 31 U.S.C. 6505), commonly known as the Thomas Amendment.

Whether the Corps should be authorized to perform NFIP levee data certifications for locally operated levees, and who would bear (or share) the costs, are matters of active debate. Some stakeholders have expressed interest in having the Thomas Amendment waived to allow the Corps to perform levee data certification. If such a change were enacted, the Corps would conduct the data certifications on a 100% reimbursable basis.

It is unknown whether the cost for the Corps to perform the certification would be less than if a private sector firm performed the certification. The Corps may be able to perform the data certification at a lower cost if it already has some of the data (e.g., for Corps-constructed projects) and if the private sector's cost is significantly influenced by liability protection.

Some stakeholders have questioned whether §5004 in WRDA 2007 (P.L. 110-114), titled "Structural Integrity Evaluations," provides the Corps with authority that could be used to undertake some of the work needed to obtain levee data certification and levee accreditation. That section included the following authority:

> Upon request of a non-Federal interest, the Secretary shall evaluate the structural integrity and effectiveness of a project for flood damage reduction and, if the Secretary determines that the project does not meet such minimum standards as the Secretary may establish and absent action by the Secretary the project will fail, the Secretary may take such action as may be necessary to restore the integrity and effectiveness of the project.

Congress has not appropriated funds for the Corps to implement this authority, and the Corps has not produced implementation guidance for this provision. The congressional reports accompanying WRDA 2007 did not provide further clarification.

There are many uncertainties about how the authority might be implemented or interpreted, including what types of projects are eligible, what types of action are authorized, what constitutes project failure, what basis the Secretary is to use for establishing minimum standards, and what the federal funding limitations may be. Therefore, the ways in which this authority might be used, especially if the provisions of the Thomas Amendment were waived for levee certification work, remain uncertain.

S. 3109 (111th Congress), the Rural Community Flood Protection Act of 2010, which is discussed in more detail in *Appendix,* would have authorized the Corps to compile the data for levee accreditation as part of its civil works program; that is, rather than the costs being reimbursable, the work would generally be cost-shared at a 65% federal and 35% nonfederal split, except for a 0% nonfederal cost share for small communities and volunteer levee operators.

Provisionally Accredited Levees

As discussed above, levee classifications on DFIRMs are determined by the FEMA levee accreditation process. The process of collecting and submitting data that document compliance with the criteria set forth in 44 C.F.R. § 65.10 may be time-consuming and expensive for communities. FEMA can offer to accredit levees provisionally while the accreditation documentation is collected and reviewed. Provisionally accredited levees (PALs) are shown on the DFIRMs as providing protection from the 1%-annual-chance event. As discussed above, the DFIRMs are used to determine flood insurance rates and purchase requirements.[34] As shown in *Table 2,* there are 295 PALs, which represent over 14% of all levees for which FEMA accreditation has been sought since mid-2006. Once a levee is provisionally accredited, the community has 24 months to submit documentation showing compliance with 44 C.F.R. § 65.10. If a community fails to produce documentation, the levee is de-accredited.

As shown in *Table 2*, as of November 2009, there were 687 levees that were provisionally accredited but for which documentation of compliance could not be produced within the required 24 months. FEMA has established guidelines for notifying communities that a PAL will be deaccredited.[35] Prior to the expiration of the PAL designation, FEMA notifies the levee system owners, the community, and other stakeholders such as congressional offices of the possible de-accreditation.

Table 2. FEMA Provisional Levee Accreditation Actions (June 2006-January 2011)

Levee Category	Number of Levee Systems[a]
Accredited[b]	3,650
Accredited, PAL[c]	1,150
De-accredited[d]	300
De-accredited, PAL[e]	687
Non-accredited[f]	8,339
Total	14,126

Source: Federal Emergency Management Agency, Mitigation Directorate, February 2011. Notes:

[a] May include non-levee embankments and similar flood structures.

[b] These levees were accredited but do not include Provisionally Accredited Levees (PALs). Therefore, the DFIRM will show the levees as providing protection from the base (1%-annual-chance) flood.

[c] These levees went through the Provisionally Accredited Levee (PAL) process and were accredited. The data and documentation indicating compliance with 44 C.F.R. § 65.10 were received and reviewed, and met the NFIP regulations. Therefore, the DFIRM will show the levees as providing protection from the base flood.

[d] These levees were de-accredited due to noncompliance with 44 C.F.R. §65.10. Therefore, the DFIRM will show the levees as not providing protection from the base flood.

[e] These levees went through the Provisionally Accredited Levee (PAL) process and were de-accredited. Either 44 C.F.R. § 65.10 data and documentation were received, but did not meet the NFIP regulations, ordata and documentation were not received to comply with 44 C.F.R. § 65.10. Therefore, the DFIRM will show the levees as not providing protection from the base flood.

[f] These levees were inventoried, but were never previously accredited or Provisionally Accredited Levee (PAL)-eligible. Therefore, the DFIRM will show the levees as not providing protection from the base flood.

Once the 24-month period expires, the PAL designation is removed, the levee system is de-accredited, and FEMA initiates a mapping project in the impacted areas. From this point, the mapping phase takes roughly 18 months; if the levee owner submits information sufficient to obtain accreditation during this period, the levee may be accredited before the updated map goes into effect. A significant portion of the recent interest in federal levee assistance has come from PAL owners struggling to assemble their accreditation packages and make levee improvements to meet FEMA's requirements. Similarly, owners of levees that have been de-accredited have shown strong interest in federal assistance for levee improvements.

LEVEE STUDY AND CONSTRUCTION

The Corps is the primary federal agency involved in the study and construction of levees. The NRCS provides limited assistance for small-scale levees. FEMA has had little involvement in levee planning and construction.

Corps Flood Damage Reduction Projects

As previously noted, the Corps, at the direction of Congress, is authorized to participate in the cost-shared planning and construction of flood damage reduction projects, such as building levees and floodwalls to reduce damages from riverine and coastal flood hazards. Corps involvement is predicated on the project being in the national interest, which is determined by the likelihood of widespread general benefits of the investment, a shortfall in the local ability to solve the water resources problem, and precedent and law.

The standard process for Corps participation in a levee project requires two separate congressional authorizations—one for investigation and one for project construction—as well as annual appropriations.[36] The investigation phase starts with Congress authorizing a study; if it is funded, the Corps then conducts an initial reconnaissance study followed by a more detailed feasibility study, which informs the congressional decision about whether to authorize the project for construction. The feasibility study analyzes whether it is in the national interest for the Corps to participate in the project and identifies the federally preferred alternative, if any. Since the mid-1980s, local project sponsors (often local governments or special levee and drainage districts) generally share construction cost of federal flood control projects and are fully responsible for their operation and maintenance. The cost-sharing for construction is 65% federal and 35% nonfederal for most flood control projects.[37] The construction cost of these projects can range from a few million dollars to more than a billion dollars. The authorized cost is generally included as part of the enacted legislative language providing congressional authorization for project construction.

NRCS Assistance for Small-Scale Levees

NRCS has authority under two programs—the Watershed and Flood Prevention Operations Program (often referred to as the Small Watershed Program)[38] and the Emergency Watershed Protection (EWP) program—to conduct small-scale levee work on private land. It is called the Small Watershed Program because no project may exceed 250,000 acres, and no structure may exceed more than 12,500 acre-feet of floodwater detention capacity, or 25,000 acre-feet of total capacity. The Small Watershed Program authorizes NRCS to provide technical and financial assistance to state and local organizations to plan and install measures to prevent erosion, sedimentation, and flood damage and to conserve, develop, and utilize land and water resources. While projects can include levees, they are on a much smaller scale compared with other federal projects.[39] Operation, maintenance, and rehabilitation of every levee built under the Small Watershed Program is the responsibility of the local project sponsor upon completion. The Small Watershed Rehabilitation Program, also administered by NRCS, is authorized to rehabilitate only dams, not levees, built under the Small Watershed Program.[40]

LEVEE RELATED FLOOD FIGHTING

The Corps and FEMA have several programs that authorize the agencies to assist with flood fighting in response to an imminent or occurring flood event. These flood fighting actions can include assisting with existing levees that are locally operated (e.g., temporary reinforcement measures) or the construction of temporary levees to manage floodwaters.

Corps Flood Emergency Response Activities

In P.L. 84-99 (33 U.S.C. § 701n), Congress gave the Corps emergency response authority that allows the agency to fight floods and other natural disasters. The law authorizes disaster preparedness, advance measures, emergency operations (disaster response and post-flood response), rehabilitation of flood control works threatened or destroyed by floods, protection or repair of federally authorized shore protection works threatened or destroyed by coastal storms, emergency dredging, and flood-related rescue operations. These activities are limited to actions to save lives and protect improved property (i.e., public facilities and services, and residential or commercial developments), and appropriations for flood related programs vary (see *Table 3*).[41]

Table 3. Appropriations for Corps Flood Control and Coastal Emergency Response (in millions)

Corps Action Under P.L. 84-99 Authority	FY2005	FY2006	FY2007	FY2008	FY2009	Total	5-Year Average
Flood fighting[a]	$32	$113	$25	$55	$50	$275	$55
RIP, Non-Katrina	$0	$57	$562	$187	$44	$850	$170
RIP, Katrina	$200	$4,828	$2,926	$0	$439	$8,393	$1,679
Total	$232	$4,998	$3,513	$242	$533	$9,518	$1,904

Source: Corps fiscal year data provided to CRS in April and May 2010.

Notes: These figures combine funding for both locally operated and federally operated levees and flood control works; data on spending only for locally operated projects are not currently available.

[a]. This includes work performed under the Corps' flood-fighting authority; it does not include the work performed as part of FEMA's emergency response activities.

FEMA's Public Assistance Program

When a disaster occurs and a state is granted federal disaster assistance under the Stafford Act, funding under the Public Assistance program may be available to reimburse communities for flood-fighting activities and emergency repairs made to eligible levees.[42] FEMA averages about $1.3 billion in PA program obligations per year.[43] Although seven categories of projects are eligible for PA funding, most levee projects are funded under the Debris Removal and Water Control Facilities categories.[44] Under the statutory provisions of the Stafford Act, the PA program does not have a funding cap, which means that all eligible projects receive federal funding for at least 75% of the project cost.[45] Regulations establish eligible levee projects under the PA program.[46] Eligible PA projects may include:

- emergency and permanent repairs to restore the levee to its pre-disaster condition;
- removal of debris in a flood control work;[47]
- placement and removal of flood-fighting measures if such activity is necessary to eliminate a public health and safety threat; and
- dewatering of areas behind the levee if there is a threat to public safety or structures.[48]

Ineligible levee activities under the PA program include:

- repair of levees that are in the Corps' Rehabilitation and Inspection Program or have a pending RIP application deemed eligible;
- repair of eligible levees under NRCS programs;
- dewatering of areas behind levees for the purpose of drying land;
- repair of secondary levees riverward of a primary levee, unless protecting human life; and
- increasing the height of a levee.[49]

Generally, PA program funds are limited to restoring a structure to its pre-disaster condition; projects to construct new levees or enhance existing levees are not eligible. In some instances, locally operated levee projects are deemed ineligible because of potential funding from other federal sources. In order to prevent duplication of programs, FEMA utilizes a decision tree for PA levee project requests, as shown in *Figure 1.*

Discretion to determine levee project eligibility for PA lies with the FEMA Administrator. Disagreement occasionally arises between FEMA officials and state officials, especially regarding the need to restore a levee to its pre-disaster condition. The Stafford Act states that federal disaster assistance, administered by FEMA, is available to restore a structure to its pre-disaster condition.[50] It can be argued that, in some cases, the most cost-effective use of federal funds would be to enhance the level of flood protection of a levee that is already undergoing repair after a disaster, rather than just restoring it to the pre-disaster level of protection.

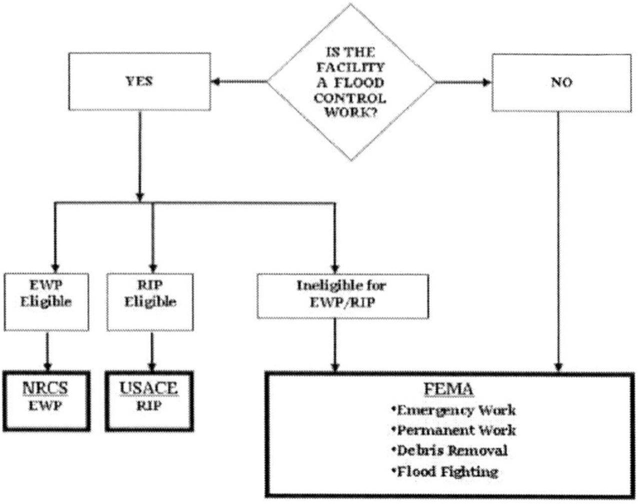

Source: U.S. Department of Homeland Security, Federal Emergency Management Agency, Rehabilitation Assistance for Levees and other Flood Control Works, Disaster Assistance Policy 9524.3, February 25, 2009.

Figure 1. Levee Eligibility for FEMA Public Assistance in Presidentially Declared Disasters.

Levee Repair, Rehabilitation, and Inspection

FEMA, the Corps, and NRCS have programs that can fund disaster related emergency levee repairs. While some of the flood fighting activities discussed above may overlap with disaster related repairs, such as the FEMA Public Assistance program, most activities covered by repair and rehabilitation assistance discussed below do not occur during a flood event. Rather, repair and rehabilitation activities are undertaken after the peak of a flood event has occurred and the extent of damage from the flood event can be determined.

Corps Rehabilitation and Inspection Program

Through its Rehabilitation and Inspection Program, the Corps provides for rehabilitation of damaged flood control works and federally constructed hurricane or shore protection projects and related inspections. To be eligible for rehabilitation assistance, the levee must be in active status at the time of the damage by wind, wave, or water action of an other than ordinary nature.[51] The following types of levees are eligible for inclusion in RIP:

- non-federally or federally constructed, locally maintained levees and floodwalls; and
- federally authorized and constructed hurricane and shore protective structures.

Eligibility is also limited to locally constructed and maintained levees must provide a minimum of a 10-year level of flood protection with two feet of freeboard to an urban area, or a minimum of a five-year level of protection with one foot of freeboard to an agricultural area. Local levee owners request that the Corps consider their levee to be included in the RIP.

To keep an active RIP status, the levee owner is required to maintain the levee properly; the sufficiency of the maintenance is determined during an annual or semi-annual inspection by the Corps, and the levee owner making taking actions to address any identified deficiencies.[52] Approximately 2,000 projects, representing 14,000 miles of levees, participate in RIP—2,250 miles of locally constructed and operated levees; 9,650 miles of Corps-constructed, locally operated levees; and 2,100 miles of Corps operated levees.[53] The Corps issued in January 2009 a policy for the temporary extension of RIP to locally operated levees with deficient conditions if the owner is undertaking system-wide improvements.[54]

For locally constructed projects, the cost to repair the damage is paid 80% by the Corps and 20% by the levee owner; for federally constructed projects, the repair cost is entirely a federal responsibility (except for the costs of obtaining the sand or other material used in the repair). For damage to be repaired, the repair must have a favorable benefit-cost ratio. Rehabilitation assistance is limited to repair or restoration of the project to its pre-disaster level of protection; no betterments or levee setbacks are allowed. Local sponsors are required to assume any rehabilitation cost of damage to an active project that is attributable to deficient maintenance. For larger floods or natural disasters resulting in the need for significant RIP-funded repairs, Congress has often used supplemental appropriations to fund eligible repairs. For smaller RIP repairs, the Corps often attempts to fund repairs within its existing funding; at times, some eligible repairs have been delayed due to limitations on the availability of funds.

A common issue that arises under RIP (as well as for FEMA mitigation programs discussed later) is interest in not only repairing levees but also improving them. Congress expressly restricted RIP funds to repair. The program is not designed to evaluate the federal interest in investments to further reduce the flood risk at a location. If federal participation is sought to increase protection, the typical route would be to pursue a study by the Corps to initiate a separate flood damage reduction project.

NRCS Emergency Repair Assistance

NRCS also has authority, under the Emergency Watershed Program, to conduct emergency levee work (among other activities) to relieve imminent hazards to life and property in the event of a natural disaster.[55] Similar to the Small Watershed Program, EWP may work only on small levees within watersheds of 400 or fewer square miles. Emergency repairs to projects designed and constructed by either the Corps or NRCS may only be made by the respective agency (i.e., NRCS cannot repair a structure built by the Corps or vice versa). For nonfederal structures, however, the division of responsibility is based on a 1986 memorandum of agreement between the Corps and NRCS.[56] Under the agreement, the Corps is responsible for repair of flood damage to projects installed for the purpose of controlling floodwaters in watersheds larger than 400 square miles. NRCS, on the other hand, is responsible for flood damage to projects built for the purpose of flood prevention in small watersheds (400 or fewer square miles). If damage is caused by a natural disaster other than a flood (e.g., fire, tornado, or earthquake) and assistance from the Corps (under P.L. 84-99) is not authorized, such assistance becomes an NRCS responsibility. Levees that qualify for EWP assistance must have a local project sponsor and benefit more than one person or interest. This restriction usually precludes assistance to many of the nonfederal levees near cropland.

Flood Hazard Mitigation

Activities that save lives and reduce damage to property are generally considered flood mitigation activities. FEMA hazard mitigation grant programs provides funding for flood mitigation activities. Hazard mitigation can reduce federal costs by decreasing the level of damage from future disasters. Since 1989, Congress has appropriated over $135.6 billion for FEMA disaster assistance programs.[57] As shown in *Table 4* and discussed in further detail below, over $8.9 billion has been made available for hazard mitigation in the last five years. Certain flood damage reduction projects are eligible under FEMA's Hazard Mitigation Grant Program (HMGP) and Pre-Disaster Mitigation grant program (PDM).

There is debate regarding the extent to which FEMA hazard mitigation assistance programs can be used for levee construction and betterment, if at all. FEMA hazard mitigation assistance programs cannot duplicate the activities of other federal agencies or be part of a larger flood control system. It is thus critical to understand the extent to which FEMA and Corps programs (and to a lesser extent NRCS programs) may or may not fund similar flood reduction projects.

Table 4. Funding for FEMA Hazard Mitigation Programs (in millions)

FEMA Program	FY2005	FY2006	FY2007	FY2008	FY2009	Total	5-Year Average
HMGP[a]	$157	$229	$269	$533	$750	$1,938	$387
PDM[b]	$100	$50	$100	$114	$90	$906	$181
PA[c]	$1,351	$2,406	$842	$427	$1,539	$6,565	$1,313
Total	$1,608	$2,685	$1,211	$1,074	$2,379	$8,957	$1,791

Source: FEMA, Office of the Chief Financial Officer, May 2010.

Notes: This list includes only FEMA hazard mitigation programs that are specifically addressed in this report. It does not include other FEMA mitigation programs, such as the Flood Mitigation Assistance, Repetitive Flood Claims, and Severe Repetitive Loss programs.

[a] Funding amounts for the Hazard Mitigation Grant Program (HMGP) are based on actual obligations by fiscal year. Appropriations for HMGP are part of a larger appropriation to the FEMA Disaster Relief Fund. Levels of funding for HMGP are determined by a formula for individual disasters. Funding levels for HMGP vary widely based on the number of FEMA disaster declarations in each fiscal year.

[b] Funding amounts for the Pre-Disaster Mitigation (PDM) program are based on congressional appropriations to the PDM account.

[c] Funding amounts for Public Assistance (PA) include Category A and Category D types of projects (see footnote 44). Many of the project categories may not provide any funding for flood control projects.

FEMA's Hazard Mitigation Grant Program[58]

The purpose of the HMGP is to reduce loss of life and property damage in future disasters. The HMGP provides grants for long-term hazard mitigation projects after a major disaster declaration. A major disaster declaration is issued by the President under the authority of the Stafford Act.[59] Once a presidential declaration has been made, hazard mitigation assistance is available and project applications can be submitted. The maximum federal cost share for hazard mitigation projects is 75%. The amount of HMGP assistance available to a state is based on a percentage of total federal disaster grants made for that declaration, with total contributions not to exceed 15% of the aggregate amount of the grants.[60]

Long-term mitigation projects may include elevating properties, acquiring properties and converting them to open space, retrofitting buildings, and constructing floodwall systems to protect critical facilities.[61] Levee repair, maintenance, and enhancement projects could possibly be considered mitigation activities and might potentially be considered eligible under the HMGP. While statutory language does not expressly prohibit funding for certain types of levees, such projects have historically not been funded under hazard mitigation assistance except when there is an immediate risk to public safety. It may be argued that failure to fund levees to enable them to maintain FEMA accreditation does create an "immediate risk to public safety." Under this view, FEMA acknowledges the public safety risk when it designates a levee for de-accreditation due to a lack of compliance with levee standards.

One possible explanation for denial of levee projects relates to the risk of duplication of programs discussed earlier in this report. Another reason may be the low prioritization of levee projects under HMGP as effective flood mitigation measures. According to the Association of State Floodplain Managers, levees should be used as a mitigation activity of last resort, because they provide limited flood protection.[62] Since communities can undertake several other flood mitigation measures, competition for flood mitigation funding may preclude funding of levee projects. Other mitigation projects that would likely receive higher prioritization than levee projects include elevation of properties, retrofitting existing structures,[63] and stormwater management.[64] One of the most effective flood mitigation activities is to buy out homes and businesses in a floodplain and convert the land to open space. Property acquisition and relocation projects are eligible under the HMGP and would most likely be funded before levee projects because removing a structure from the floodplain is more effective in reducing flood damages.

FEMA's Pre-Disaster Mitigation Grant Program[65]

The Pre-Disaster Mitigation (PDM) program funds structural and nonstructural mitigation projects prior to a disaster occurring in a community.[66] Eligible projects are similar to those funded under the HMGP: property acquisition and relocation, structural elevation, and other, non-flood related mitigation. There is a 25% state/local cost-share requirement for PDM grants. Because of the costly nature of levee projects, the cost-share provision may make the PDM program a less viable option than HMGP for communities considering such projects, assuming the other barriers to using HMGP, such as duplication of programs, can be overcome. As shown in *Table 5,* the PDM state/local cost-share is still lower than the Corps' Flood and Coastal Storm Damage Reduction Studies and Construction program.

Comparison of FEMA, Corps, and NRCS Levee Programs and Authorities

Table 5 summarizes information on the Corps, FEMA, and NRCS programs previously described; it presents an overview of the federal programs that are most often discussed as potential sources of funds for locally operated levees. The table clarifies that the Corps is the main federal partner in the construction of locally operated levees. What it does not show is that, as previously discussed, the process for obtaining this federal construction assistance is a lengthy multi-step process, and there is significant competition for available federal funds among congressionally authorized projects. *Table 5* shows that there are no general federal authorities for assistance with the regular operation and maintenance of locally operated levees. It further shows that there are multiple possible authorities for federal assistance for levees damaged by natural events and that each authority has its own cost share; which of these authorities is appropriate depends on the type of event (e.g., declared disaster) and the type of levee (e.g., small).

Table 5. Role of Locally Operated Levees under Select Corps, FEMA, and NRCS Programs and Authorities

Federal Program or Activity	Eligible Locally Operated Levee Projects	Disaster-Related Timing of Levee Work	Max. Federal Cost Share	Type of Levee Work Eligible	Significance of Levees' Current Role in Program	Action Required for Levee Project to Access Program	Avg. Program Funding for FY2005-FY2009(in millions)
Corps—Flood and Coastal Storm Damage Reduction Studies and Construction	Congressionally authorized[a]	None	65%	Design and construction	Significant, but not the only work funded	Congressional study and construction authorization and, annual congressional appropriations	$857[b]
Corps—Operation and Maintenance[c]	None	—	—	—	—	—	—
Corps—Emergency Flood Fighting	Levees protecting life and improved property	Imminent disaster or during disaster	100%	Emergency, temporary strengthening	Significant, but not the only work funded	Governor or nonfederal project sponsor requests assistance; Corps uses its discretion to perform the work deemed necessary and within its authority	$55
Corps—RIP	RIP active levees and well-maintained levees	Post-disaster	100% for federally constructed, 80% for locally constructed	Repair of damage from a natural event	Significant, majority of work funded	Owner applies; Corps evaluates the eligibility and performs the work when funds become available	$1,849

Federal Program or Activity	Eligible Locally Operated Levee Projects	Disaster-Related Timing of Levee Work	Max. Federal Cost Share	Type of Levee Work Eligible	Significance of Levees' Current Role in Program	Action Required for Levee Project to Access Program	Avg. Program Funding for FY2005-FY2009(in millions)
FEMA—HMGP	Projects ineligible under Corps and NRCS programs	After a declared disaster	75%	Design and construction projects ineligible under Corps or NRCS programs	Insignificant, usually not eligible	FEMA approves and selects from eligible projects based on prioritization criteria	$387
FEMA—PDM	Projects ineligible under Corps and NRCS programs	Before a declared disaster	75%[d]	Design and construction projects ineligible under Corps or NRCS programs	Insignificant, usually not eligible	FEMA approves and selects from eligible projects based on prioritization criteria	$181
FEMA—PA	Projects ineligible under Corps and NRCS programs	After a declared disaster	75%	Emergency repairs ineligible under Corps or NRCS authorities	Insignificant, small percentage of project funded	FEMA approves and selects from eligible projects based on prioritization criteria	$1,313
NRCS—Small Watershed Program	Small levee projects as part of small watershed protection	None	100%	Design and construction	Insignificant, projects rarely funded	Selected by NRCS from eligible projects based on prioritization criteria	$58

Table 5. (Continued)

Federal Program or Activity	Eligible Locally Operated Levee Projects	Disaster-Related Timing of Levee Work	Max. Federal Cost Share	Type of Levee Work Eligible	Significance of Levees' Current Role in Program	Action Required for Levee Project to Access Program	Avg. Program Funding for FY2005-FY2009(in millions)
NRCS—Emergency Watershed Program	Small levee projects that reduce threats to life and property	Post-disaster[e]	75%	Repair of damage from a natural event	Insignificant, projects rarely funded	Owner applies; NRCS distributes funds to eligible projects	$191

Source: Congressional Research Service.

[a]. The Corps also has limited authorities and funding to undertake small levee projects (less than $7 million in federal funds) without project-specific congressional authorization.

[b]. This figure represents all Corps flood and coastal storm damage reduction funding through annual appropriations, except for the Mississippi River and Tributaries budget account. (The Corps funding for the Mississippi River and Tributaries projects could not be calculated because the data do not include supplemental appropriations.) This figure includes the funds for projects that would be operated by the Corps and by local levee owners.

[c]. Although the Corps lacks general authority and funding to perform maintenance of locally owned levees, the agency received on average $632 million annually between FY2005 and FY2009 for operation and maintenance of Corps-operated levees and for its Mississippi Rivers and Tributaries Project.

[d.] There is a maximum federal cost share of 90% for communities meeting the eligibility criteria for small and impoverished communities.

[e.] The levee owner must apply to NRCS within 10 days after the disaster for exigency situations and within 60 days for nonexigency situations.

Policy Options for Assisting Locally Operated Levees

As previously noted, Congress is debating whether and how to change the current division of levee responsibilities and their costs. Some stakeholders seek expansion of activities eligible for federal assistance, while others are concerned with the federal government assuming more of the cost and liability for levee investments that they consider a local responsibility. This section discusses some of the options available for congressional consideration. These include, but are not limited to, maintaining the status quo, adopting the recommendations of the congressionally established National Committee on Levee Safety, creating a new grant program, supplementing existing grant programs, or reducing the federal role in levee funding.

Maintain the Status Quo

As discussed in this report, there is debate about whether existing federal programs align with the appropriate federal role and whether such programs adequately address locally operated levee needs. One driver prompting the maintenance of the status quo is that most of the alternatives being discussed likely would increase federal responsibility, might increase federal liability concerns, and would require additional federal funding to implement.

Implement Recommendations of the National Committee on Levee Safety

In § 9003 of the Water Resource Development Act of 2007 (P.L. 110-114), Congress created a National Committee on Levee Safety (NCLS) to study and make recommendations for a national levee safety program. The NCLS is comprised of sixteen members, each with expertise in some aspect of levee safety. One member is from the Corps, one from FEMA, eight from state levee safety agencies, two from the private sector, two from local/regional governments, and two from Indian tribes. In 2009, the NCLS completed a draft report, including 20 recommendations embracing three main concepts:

- the need for national leadership through a National Levee Safety Commission that would advise on national technical standards, risk communication, and coordination of environmental and safety concerns,
- the creation of a levee safety program in all states with responsibilities for oversight and regulation, and
- the need for an alignment of federal agency programs.[67]

Legislation is needed to implement 12 of the 20 recommendations fully and nationally. The report remains in draft form; that is, it has not been formally transmitted from the Administration to Congress. The House Transportation and Infrastructure Committee, Subcommittee on Water Resources and the Environment, received testimony on the recommendations in May 2009.[68]

The NCLS contends that states, not the federal government, should have primary authority for implementation of a national levee safety program. Ten states keep a list of levees within their borders, and 23 states have an agency with some levee safety responsibilities. The recommendation is to create a new commission that will establish a national levee safety standard for use by state programs. The NCLS draft proposes, among other investments:

- a new levee safety grant program to assist states in achieving strong levee safety programs, at $113 million annually in federal appropriations;
- a National Levee Rehabilitation, Improvement, and Flood Mitigation Fund to address both structural and nonstructural levee rehabilitation needs,[69] at $600 million annually in federal appropriations; and
- authority for the Corps to perform a one-time inspection of all locally operated levees (not only federally constructed levees or those participating in RIP) to support the development of the National Levee Database, at $125 million annually for the next five years.[70]

The cost to implement the full suite of recommendations in the report for the first five years would amount to $1.238 billion annually—$878 million in federal funds and $360 million in nonfederal funds according to the draft report. Most of the federal funding would be directed toward new federal activities—that is, activities that the federal government generally does not currently fund.

Elements of the legislation discussed in the next section of this report and *Appendix* may run counter to the NCLS recommendations. There has not been uniform acceptance of the draft recommendations; for instance, some view implementation of the recommendations as creating an unfunded mandate for levee safety that falls too heavily on the states and levee owners. Others are concerned about the creation of a new bureaucratic entity.

Create a New Program

Another option is to create a new grant program that would provide targeted funding to assist with aspects of locally operated levees that are not addressed by current programs (see *Table 5).* The new grant program would be distinct from the new levee safety program recommended by the NCLS because the project eligibility could be more extensive. Creating any new program requires many policy decisions to define such criteria as eligible activities, prioritization of activities, funding sources, nonfederal cost shares, the party responsible for performing the work, and the form of assistance (e.g., loan or grant). An additional consideration is where to house the program; Congress may consider the advantages and disadvantages of selecting various agencies to administer the new program.

Supplement Appropriations for Existing Programs

Congress could expand existing programs, either by expanding their authorization or by increasing appropriations. For example, FEMA arguably has the authority to fund locally operated levee repair and rehabilitation projects when there is no risk of duplicating benefits or activities of other federal agencies.[71] Where program duplication is unclear, a memorandum of understanding with appropriate agencies would assist the agency personnel in making eligibility determinations. However, as discussed in this report, funding is limited under existing programs. Congress may consider supplementing existing FEMA or Corps programs to provide additional funding for locally operated levee projects.

Reduce the Federal Role

Congress may assess whether the current federal role in locally operated levees is justified, in particular whether the current federal role provides incentives for floodplain use consistent with the national interest. Current federal programs are often rooted in historical concerns about public safety and reducing property damage. Local governments are traditionally responsible for activities such as land use and zoning, which influence flood risk. States also influence floodplain development decisions in an oversight role. It could be argued that if local and state governments choose to allow development in a floodplain, the cost of leveraging the flood risk to structures through structural flood control projects such as levees should fall exclusively on the local and state governments. Congress may consider reducing the federal role in locally operated levees by reducing appropriations for programs that fund such levees.

LEGISLATIVE DEVELOPMENTS IN THE 112TH CONGRESS

The 112th Congress has expressed interest in the mapping methodology used by FEMA in updating the FIRMs. On February 3, 2011, 27 U.S. Senators sent a letter to FEMA Administrator Craig Fugate requesting that FEMA discontinue its "without levee" analysis for determining FIRMs for areas where existing levees have not obtained FEMA accreditation if the community in the affected area objects.[72] Members of Congress also have introduced legislation to address issues with updates to FIRMs and the impact of new maps on the purchase requirements for the NFIP. In addition to the legislation discussed below, efforts to address the challenges facing locally owned levees may be pursued through efforts to reauthorize the NFIP. Addressing local levee issues in a reauthorization or other bill raises a budgetary consideration. Increases in the authorizations or appropriations in federal assistance for locally owned levees would increase the demands on federal fiscal resources. Congress, the Administration, and the public are carefully scrutinizing actions that would expand federal activities given the current federal fiscal climate.

Legislation on NFIP Mapping and Flood Insurance

H.R. 764, Fair Treatment of Existing Levees Act of 2011

H.R. 764 would prohibit the FEMA Administrator from assuming that an existing levee or flood FIRM for an area if it results in the area having a new flood hazard designation. The bill would provide an exception to the prohibition if no affected community has notified FEMA within 90 days of enactment of this act of objections to the hazard modeling processes, and provided that the affected communities have provided sufficient notification to affected residents prior to the implementation of the provision.

H.R. 898, To Suspend Flood Insurance Rate Map Updates in Geographic Areas in which Certain Levees Are Being Repaired

H.R. 898 would prohibit the FEMA Administrator from updating the floodplain designation or flood risk zone for areas behind levees determined by the Corps to have a design deficiency and where the responsible entities have implemented a plan that meets certain criteria to repair the levee. In addition, state and local governments with jurisdiction over the area must have implemented a surveillance and operations plan, an evacuation plan, and an outreach and communication plan, except where the flood protection system provides protection from the 1% annual chance flood or where the FEMA administrator has suspended the updates to the floodplain areas or flood risk areas for seven consecutive years.

CONCLUDING REMARKS

While considering NFIP reauthorization, which expires September 30, 2011, and during other legislative deliberations, Congress may decide whether, and if so how, to address locally operated levees having difficulty obtaining and maintaining FEMA accreditation, and whether to change the current federal assistance provided to locally operated levees. Congressional action may clarify federal authorities for activities assisting locally operated levee and may reduce the risk of duplication of programs. Although levee issues have become a more prominent subject of congressional interest in recent years, consensus on how federal policy, programs, and funding should evolve remains elusive.

Appendix. Levee Legislation Considered in the 111th Congress

Legislation on Levee Accreditation

Two bills related to levee accreditation were introduced in the 111th Congress. In a March 2010 letter to the FEMA Administrator and the Assistant Secretary of the Army for Civil Works, 16 Senators in the 111th Congress requested a meeting to discuss levee accreditation and flood mapping processes.[73] Also during the 111th Congress, roughly 40 Members were part of the Congressional Levee Caucus; the Caucus chair announced levee certification as its chief priority.

H.R. 4935, PAL Extension

Referred to the House Committee on Financial Services, H.R. 4935 would have provided the FEMA regional office directors with the authority to grant a PAL extension not to exceed 24 months for communities making a "good faith effort" to comply with the FEMA levee accreditation requirements. The good faith effort would include documentation that a community has adequate funding for levee rehabilitation and has retained a private contractor or appropriate federal agency to verify the certification of the levee.

S. 3109, Rural Community Flood Protection Act of 2010

Referred to the Senate Committee on Environment and Public Works, S. 3109 would have authorized the Corps to perform NFIP data certification as part of the agency's civil works program rather than as a reimbursable activity. The certification would be cost-shared, with a 65% federal and 35% local split, except that nonfederal interests representing fewer than 10,000 people and volunteer levee operators would have no local cost-share. The bill would not have authorized a specific level of appropriations. Data on the average costs of data certification are not currently available.

The bill also would have addressed levee owners' concerns about the financial burden of obtaining data certification by authorizing the Corps to perform at full or partial federal expense work that under the NFIP is considered the levee owners' responsibility. Under the provisions of the bill, the federal government would assume all of the cost associated with levees protecting smaller communities and run by volunteers; most likely, these would be in rural areas. It is unclear how many such levees exist, since few rural levees provide protection from a 100-year flood. Furthermore, it is unclear whether such rural levees represent the greatest risk to life and property and therefore should receive a greater share of federal investment. The greater federal cost share in the bill for rural levees appeared to be based on a perception of a lower ability among these levee owners to pay for data certification, rather than on a greater federal stake in these levees.

Legislation on NFIP Mapping and Flood Insurance

S. 3285, Suspension of Flood Map Modernization Updates

Referred to the Senate Committee on Banking, Housing, and Urban Affairs, S. 3285 would have required that the FEMA Administrator suspend any map updates under Map Mod for seven years for counties with a flood protection structure built or maintained by the Corps. The proposed legislation would not have suspended map updates for counties protected by locally operated levees.

H.R. 3415 and S. 3051, Suspension of Flood Insurance Rate Maps in Areas where Certain Levees Are Being Repaired

H.R. 3415 and S. 3051 are identical bills that would have suspended flood insurance rate map updates for up to seven consecutive years in communities that contain levees with a design deficiency, if the community has a repair plan that was developed by the Corps or a licensed professional engineer, was approved by FEMA or the Corps, is based on reasonably current design data, and includes an adequate financing mechanism for implementing the plan. The repair plan must also ensure that once the repairs are completed, the levee would provide protection from a 100-year flood. H.R. 3415 was referred to the House Committee on Financial Services, and S. 3051 was referred to the Senate Committee on Banking, Housing, and Urban Affairs.

Legislation on FEMA's Hazard Mitigation Programs

H.R. 1746 (P.L. 111-351) and H.R. 3377, Authorizations for the Pre-Disaster Mitigation Grant Program

Passed by the House and the Senate in December 2010, H.R. 1746 (P.L. 111-351) reauthorized the PDM program for three fiscal years beginning in FY2011, with authorization for $180 million in fiscal year 2011, $200 million for fiscal year 2012, and $200 million in FY2013. The legislation increased the minimum award amount for each state by providing the lesser of $575,000 or 1% of total program appropriations, with a maximum per state award not to exceed 15% of the fiscal year appropriation. The enacted legislation also included a prohibition on congressionally directed spending for the PDM program.

While H.R. 3377 would have authorized an annual appropriation of $250 million for three years, beginning in FY2010, the bill did not receive further consideration because H.R. 1746 was passed by both the House and the Senate.[74] The PDM program is a potential source of funding to address the levee issues presented in this report. Appropriations for PDM have fluctuated significantly since initial program authorization, ranging from $150 million for FY2003 and FY2004 to $50 million in FY2006.

End Notes

[1] For more on the authorization of the National Flood Insurance Program, see CRS Report R40650, *National Flood Insurance Program: Background, Challenges, and Financial Status* , by Rawle O. King.

[2] Local levee owners can include a wide variety of stakeholders. There are private levees (e.g., individual farmers, businesses, and homeowners) and publicly owned levees (e.g., levee districts, water control or improvement districts, municipalities, and states). Anyone can own a levee.

[3] The operation and maintenance of levees generally involves mowing the grass on and around the levee, removing trees, and repairing damage done by animals.

[4] In this report, the term *levees* is used broadly to encompass both levees and floodwalls. Levees often are broad, earthen structures, while floodwalls are concrete and steel walls, built atop a levee or in lieu of a levee. Floodwalls are often used in urban areas because they require less land than levees.

[5] National Committee on Levee Safety, *DRAFT Recommendation for a National Levee Safety Program*, January 2009, p. 13, available at http://www.iwr.usace.army.mil/ncls/docs/NCLS-Recommendation-Report_ 012009_ DRAFT.pdf.

[6] Other federal entities also own and maintain some levees or structures that at times may be viewed as functioning like levees. These include the International Boundary and Water Commission (IBWC), the Tennessee Valley Authority (TVA), and the Department of the Interior's Bureau of Reclamation (Reclamation). The IBWC operates roughly 610 miles of levee (IBWC, *Rio Grande Flood Control System Rehabilitation Project*, http://www.ibwc.gov/Recovery/RGF.html, visited March 2011). No official estimates of levee miles operated by TVA or Reclamation are avilable.

[7] American Society of Civil Engineers, *2009 Report Card for America's Infrastructure*, March 2009, p. 7, available at http://www. infrastructurereportcard.org/.

[8] The Corps published a *Design and Construction of Levees Engineering Manual*, which is currently in the process of being updated. The current version from April 2000 is available

at http://140.194.76.129/publications/eng-manuals/em1110-2-1913/toc.htm. The manual presents basic principles of earthen levee designs to be used as a guide for Corps engineers, rather than as a code or minimum design standard for the broader engineering community.

[9] 42 Stat. 596.

[10] Jamie W. Moore and Dorothy P. Moore, *The Army Corps of Engineers and the Evolution of Federal Flood Plain Management Policy* (Boulder, CO: University of Colorado, 1989), pp. 8-13.

[11] P.L. 90-448. The NFIP was broadened and modified with the passage of the Flood Disaster Protection Act of 1973 (P.L. 93-234), and other major reform legislation in 1977 (P.L. 95-128), 1994 (P.L. 103-325), and 2004 (P.L. 108-264).

[12] Among them, for example, are regulatory entities such as the Tennessee Valley Authority, the Bureau of Reclamation, the International Boundary and Water Commission, and agencies such as the U.S. Geological Survey that provide data used in assessing flood risk.

[13] U.S. Department of Homeland Security, Federal Emergency Management Agency, *Rehabilitation Assistance for Levees and Other Flood Control Works*, Disaster Assistance Policy 9524.3, February 25, 2009.

[14] 44 C.F.R. § 206.434(g).

[15] FEMA *Rehabilitation Assistance for Levees and Other Flood Control Works*, Disaster Assistance Policy 9524.3, February 25, 2009.

[16] E-mail correspondence between Natalie Keegan and FEMA Office of Legislative Affairs, May 4, 2010.

[17] Telephone conversation between Natalie Keegan and Keith Turi, FEMA Mitigation Directorate, and Vince Fabrizio, FEMA Legislative Affairs, September 9, 2008. For additional information on use of CDBG funds in disasters, see CRS Report RL33330, *Community Development Block Grant Funds in Disaster Relief and Recovery*, by Eugene Boyd.

[18] Corps maintained levees are primarily those along the Mississippi River and those that the Corps constructed prior to the early 1970s. It also operates numerous multi-purpose dams that provide flood damage reduction benefits.

[19] U.S. Army Corps of Engineers, *USACE Process for the National Flood Insurance Program (NFIP) Levee System Evaluations*, EC 1110-2-6067, Washington, DC, August 31, 2010, http://140.194.76.129/publications/ eng-circulars/ec1110-2-6067/.

[20] See *Paterno v. State of California*, 2003 Cal. App. LEXIS 1771 (2003), pet. for rev. denied, 2004 Cal. LEXIS 2253 (March 17, 2004); see also *Arreola v. County of Monterey*, 2002 Cal. App. LEXIS 4319 (2002), pet. for rev. denied, 2002 Cal. LEXIS 6194 (September 18, 2002).

[21] The issue of federal liability for damages is discussed in CRS Report RL34131, *Federal Liability for Flood Damage Related to Army Corps of Engineers Projects*, by Cynthia Brougher.

[22] For additional information on levee liability issues, see CRS Report 95-717, *Federal Tort Claims Act (FTCA)*, by Vivian S. Chu.

[23] Ibid., pp. 8-11.

[24] See 42 U.S.C. 4012(c), 4022, and 4102.

[25] For more information and an evaluation of the current process, U.S. Government Accountability Office, *FEMA Flood Maps: Some Standards and Processes in Place to Promote Map Accuracy and Outreach, but Opportunities Exist to Address Implementation Challenges*, GAO-11-17, Dec. 2010.

[26] U.S. Department of Homeland Security, Federal Emergency Management Agency, email correspondence from Vince Fabrizio, FEMA Legislative Affairs, to Natalie Keegan, Congressional Research Service, Feb. 2, 2011. The approximate number of levees combines accredited levees and provisionally accredited levees (PALs), as detailed in Table 2.

[27] See http://www.fema.gov/library/viewRecord.do?id=2367.

[28] The freeboard is the distance between normal water level and the top of a structure.

[29] 44 C.F.R. § 65.2(b).

[30] Supporting data could include analyzing samples of levee construction materials and modeling flood flows. Improvements may include increasing levee height, hardening the levee, protecting the base (also known as the toe) of the levee, and removing encroachments 31 44 C.F.R. § 65.10(b).

[32] National Committee on Levee Safety, *Draft: Recommendations for a National Levee Safety* Program, January 15, 2010, p. 90, at http://www.nfrmp.us/ncls/docs/NCLS-Recommendation-Report_012009_ DRAFT.pdf. The inspections performed by the Corps for inclusion and active status in RIP do not include engineering assessments for project performance or stability, which are among the factors evaluated by FEMA when accrediting a levee.

[33] See endnote 19

[35] U.S. Department of Homeland Security, Federal Emergency Management Agency, *Procedural memorandum 53: Guidance for Notification and Mapping of Expiring Provisionally Accredited Levee Designations*, April 24, 2009.

[36] One exception is for small projects. Congress has given the Corps limited authorities to undertake small projects (e.g., less than $7.0 million) without project-specific congressional authorization. For more information, see Corps' Continuing Authorities Programs in CRS Report R41243, *Army Corps of Engineers Water Resource Projects: Authorization and Appropriations*, by Nicole T. Carter and Charles V. Stern.

[37] For more on Corps flood control studies and construction projects, see CRS Report R41243, *Army Corps of Engineers Water Resource Projects: Authorization and Appropriations*, by Nicole T. Carter and Charles V. Stern.

[38] The Small Watershed Program is authorized by P.L. 83-566 and P.L. 78-534. Both authorities have similar objectives and are implemented following similar procedures. The vast majority of the projects have been built under the authority of P.L. 83-566, the Watershed Prevention and Flood Protection Act of 1954. Only 11 projects are specifically authorized under the Flood Control Act of 1944, and are much larger and more expensive than P.L. 83-566 projects.

[39] For more information, see CRS Report RL30478, *Federally Supported Water Supply and Wastewater Treatment Programs*, coordinated by Claudia Copeland.

[40] According to 16 U.S.C. § 1012(a)(3), only a structural measure defined as "a physical improvement that impounds water, commonly known as a dam, which was constructed as part of a covered water resource project, including the impoundment area and flood pool" is eligible for the Small Watershed Rehabilitation Program.

[41] Although the Corps' account paying for these activities may receive some appropriations in the annual Energy and Water Development appropriations acts, the initial appropriation is often supplemented with emergency appropriations specific to the emergency being addressed.

[42] 42 U.S.C. § 5172(b).

[43] U.S. Department of Homeland Security, Federal Emergency Management Agency, *Average Total Obligations by Year and by Declaration*, January 2009, available at http://www.fema.gov/government/grant/pa/stat2.shtm.

[44] 44 C.F.R. § 206.226(a). The public assistance categories include Category A (Debris Removal), Category B (Emergency Protective Measures), Category C (Roads and Bridges), Category D (Water Control Facilities), Category E (Buildings and Equipment), Category F (Utilities), and Category G (Parks, Recreational, and Other).

[45] 42 U.S.C. § 5172(b).

[46] 44 C.F.R., §206.220-206.228.

[47] The debris must be the result of the disaster and must pose an immediate threat to life, public health, or property.

[48] U.S. Department of Homeland Security, Federal Emergency Management Agency, *Rehabilitation Assistance for Levees and other Flood Control Works*, Disaster Assistance Policy 9524.3, February 25, 2009. Eligible dewatering activities may include a deliberate breach or pumping.

[49] The regulations provide only a general prohibition of increasing the height of a levee under the PA program eligibility guidelines. It would appear that the policy allows the FEMA Administrator some discretion in approving levee projects that would increase the height of eligible levees.

[50] 42 U.S.C. § 5172(e).

[51] 33 U.S.C. 701n. For more information on RIP, see USACE, Engineer Regulation 500-1-1, *Emergency Employment of Army and Other Resources Civil Emergency Management Program*, available at http://140.194.76.129/publications/eng-regs/.

[52] An aspect of RIP implementation receiving attention is the Corps' guidance on vegetation on levees. Some levee owners are having difficulty conducting regular maintenance and emergency repairs while also complying with environmental laws, such as the Endangered Species Act. In some areas, the vegetation on and near levees provides species habitat and other environmental benefits. This and other environmental issues associated with levee maintenance are beyond the scope of this report.

[53] Corps data provided to CRS on April 30, 2010.

[54] The policy is available at http://www.iwr.usace.army.mil/nfrmp/docs/HQS-ECOPY3I50-Exchange-01132009-162045.pdf.

[55] Authorized under Section 216, P.L. 81-516 (33 U.S.C. § 701b1) and Sections 403-405, P.L. 95-334 (16 U.S.C. §§ 2203-2205).

[56] *Memorandum of Agreement Between the Soil Conservation Service, Department of Agriculture and the Department of the Army*, May 20, 1986, signed by Robert K. Dawson, Assistant Secretary of the Army and Wilson Scaling, Chief of the Soil Conservation Service.

[57] For additional information on FEMA disaster program appropriations, see CRS Report R40708, *Disaster Relief Funding and Emergency Supplemental Appropriations*, by Bruce R. Lindsay and Justin Murray.

[58] For more information, see CRS Report R40471, *FEMA's Hazard Mitigation Grant Program: Overview and Issues*, by Natalie Keegan.

[59] 42 U.S.C. § 5170. For additional information on disaster declarations, see CRS Report RL34146, *FEMA's Disaster Declaration Process: A Primer*, by Francis X. McCarthy.

[60] 42 U.S.C. § 5170c(a). HMGP assistance shall not exceed 15% for amounts up to $2 billion, 10% for amounts between $2 billion and $10 billion, and 7.5% for amounts more than $10 billion but not more than $35.3 billion total (less any associated administrative costs).

[61] U.S. Department of Homeland Security, Federal Emergency Management Agency, *Fact Sheet: Hazard Mitigation Grant Program* (Washington, DC, June 2007), p. 1.

[62] Association of State Floodplain Managers, *National Flood Policy Challenges: Levees, The Double-Edged Sword*, white paper, April 17, 2007.

[63] Retrofitting existing buildings can include activities such as adding shutters to protect against high winds, attaching reinforcing clips to roofing, or using earthquake-resilient building materials to replace older portions of a structure.

[64] U.S. Department of Homeland Security, Federal Emergency Management Agency, *Hazard Mitigation Grant Program Desk Reference* (Washington, DC, October 1999), p. 7-1.

[65] For more information, see CRS Report RL34537, *FEMA's Pre-Disaster Mitigation Program: Overview and Issues*, by Francis X. McCarthy and Natalie Keegan.

[66] 42 U.S.C. § 5170(c).

[67] National Committee on Levee Safety, *Draft: Recommendations for a National Levee Safety* Program, January 15, 2010, at http://www.nfrmp.us/ncls/docs/NCLS-Recommendation-Report_ 012009_ DRAFT. pdf. The NCLS uses the term "alignment" without providing clarity on which programs and agencies would be involved and what is meant by "alignment."

[68] The video of the May 19, 2009 hearing is available at http://transportation.house.gov/ hearings/hearingDetail. aspx?NewsID=887.

[69] The fund is intended for locally operated levees for pre-disaster funding; it is not intended to replace FEMA's mitigation programs.

[70] The database and NCLS authority are currently capped at $20 million annually from 2008 to 2013.

[71] 44 C.F.R., §206.434(g).

[72] U.S. Senate, letter to FEMA Administrator Craig Fugate, Feb. 3, 2011, available from the authors.

[73] The text of the letter is available at http://durbin.senate.Gov/ showRelease.cfm? releaseId = 323237.

[74] In December 2010, H.R. 1746 became public law (P.L. 111-351).

INDEX

G

H

I

J

L

M

N

O

P

R